TOWNS AND CITIES: FUNCTION IN FORM

Towns and Cities: Function in Form

Urban Structures, Economics and Society

JULIAN HART
Managing Director, Lancefield Consulting Ltd, UK

Routledge
Taylor & Francis Group
LONDON AND NEW YORK

First published 2015 by Ashgate Publishing

Published 2016 by Routledge
2 Park Square, Milton Park, Abingdon, Oxon OX14 4RN
605 Third Avenue, New York, NY 10017

First issued in paperback 2021

Routledge is an imprint of the Taylor & Francis Group, an informa business

British Library Cataloguing in Publication Data
A catalogue record for this book is available from the British Library

The Library of Congress has cataloged the printed edition as follows:
Hart, Julian.
Towns and cities, function in form : urban structures, economics and society / by Julian Hart.
pages cm. -- (Design and the built environment)
Includes bibliographical references and index.
ISBN 978-1-4724-5855-1 (hardback)
1. City planning. 2. Cities and towns. 3. Urbanization. I. Title.
HT166.H3874 2015
307.76--dc23

2015010997

ISBN 13: 978-0-367-78732-5 (pbk)
ISBN 13: 978-1-4724-5855-1 (hbk)

Contents

List of Figures and Tables

Figures

Tables

List of Boxes

About the Author

Julian Hart is a leading urban practitioner operating primarily in London, specialising in programme and project delivery of larger urban/estate regeneration and housing projects. He has worked on several major development and estate regeneration schemes across the capital, usually in a client or project delivery role. Throughout his career he has worked at the leading edge of best practice, innovating approaches and techniques for delivering new development. He is also a recognised expert on urban design and housing design; he was responsible for bringing into effect the London Housing Design Standards, has sat on various design review panels and co-authored a publication championing the new London vernacular for housing.

Foreword – A Town Planning Perspective

John Rhodes, Founder and Director, Quod Limited
(Planning Consultancy)

Urban design – in its broadest sense (that is the design of towns and cities) – is an undeveloped science. Whilst it tends to be the domain of architects, as this book reminds us there is so much more to it than design in the traditional sense. To attempt to plan and construct environments for modern living requires an understanding that transcends one discipline – the designer needs a robust understanding of economics and all of the social sciences (including psychology and philosophy) as well as an engineer's toolbox and an artist's heart. The designer also needs a clear vision of what one should be trying to achieve – what is the goal, what is good and how can it be recognised, let alone delivered?

This book dares to ask "why?" and to point out how seldom that question is asked. The book challenges our evolving understanding of how to create towns and cities. Anyone involved in large scale urban development projects would benefit from being challenged by the book's assessment of traditional values. As just one example: do we truly understand how, when and why a road becomes a street? Towns need roads for the rapid movement of goods and people – free flowing roads serve an efficient economy and allow towns to feed themselves and to trade but they can be characterless, dividers of communities. As they approach their destination – often a town centre, their role and form changes – congestion or at least bustle can be good, interchange and interaction encouraged by street crossings, intersections, pedestrian activity and good design themselves become important for trade but also for social interaction. The designer needs to understand the role of both, how to plan for both and what the transition of a road to a street means for urban design, for density, for the presence and design quality of frontage buildings and for the efficient distribution of land uses. The designer of a street needs to be so much more than a traffic engineer.

The book breaks down the urban form into its constituent parts and seeks a proper understanding of the needs and function of each part – from house design to public spaces, to town centres and central business districts (what truly makes them tick?), before reviewing how they might be reassembled. Established thinking is challenged – does increasing density really reduce energy consumption by aggregating activity when allowance is made for the energy cost of vertical travel? Do higher densities encourage or undermine the potential for social interaction? Does the theory of aspiring to achieve ever increasing levels of accessibility break down when faced with the very human desire for privacy? Every urban planner and designer should allow themselves to be provoked by these challenges.

The book examines lessons from history but also asks whether our traditional understanding of central place and other urban economic theories has been ripped apart by the impact of the car and the internet. Some questions are timeless but all are timely as we embark on a renewed enthusiasm for new market towns and garden cities, partly as a response to the disappointing experience of many urban extensions. Planners and designers are once again playing God and it is essential that they are as well-equipped as they might be for the task. Indeed, as the book points out, the apparently utopian world of market towns is itself threatened by strong economic forces, with many struggling to maintain the strength to command a sufficient gravity over their population.

Lessons must constantly be learned. For instance, the book criticises an excess of out of town shopping, poor engineering based transport planning and a failure to prevent low density anonymity. I might argue that the answer to many failures of the past is not stronger negative planning policy to preserve the status quo but a more dynamic approach to positive planning. All urban structures need to be flexible and adaptable to change – our powers as planners are limited in the face of wider economic forces and we should avoid either and attempt to wind back the clock or the fallacy of assuming that we alone can command the future. Good town planning needs to embrace change and to work with rather than against market forces if it is to create new durable places which stand the test of time. The best developers know the economics of good development but also the value of good design. The best designs embrace market economics and then create social inclusion, attractive environments and secure communities in inherently sustainable locations – all attributes which attract and which, therefore, command a price premium. Confident, positive planning based on a thorough understanding of these drivers is essential if the next phase of urban development is to avoid its own negative reviews.

The book makes an important contribution to our understanding of what makes a successful place – but it does not propose a blueprint for every new community. Thankfully, each place is different – differences derive from accidents of history, the uniqueness of landscape, the economic interaction with a hierarchy of towns, communications, climate and many more distinguishing characteristics. However, we do need the best understanding that we can achieve of those matters that we can influence through layout, design and other interventions. This book tests our understanding and warrants its place alongside the best design guides as a necessary prompt to question accepted solutions.

Foreword – A Design Perspective

Professor Peter Bishop, University College London and Director of Allies and Morrison

We are in a period of rapid urbanisation, without parallel in human history. We are consequently building and renewing urban fabric at a pace that often fails to allow for the depth of design consideration that is required. We are also in the midst of technological, societal, economic and political change that is fundamentally changing our expectations of, and relationship with, the cities and towns that we live in. And on top of this our current environmental impact is unsustainable. To compound this, all of this is happening in the absence of the theory and tools that can bridge the gap between the technological possibilities open to us as designers and the expectations of citizens who will inhabit and use cities that we create. Unless we can resolve this conundrum we face a dysfunctional urban future.

The specific disciplines of town planning and urban design are still relatively young, the product of, and response to, late 19th century and 20th century urbanisation. The complexity of social, political and economic issues, together with the scale of urban growth, outstripped the ability of traditional design practices, largely focused on the articulation of physical form, to adequately respond. Although the successes of 20th century urban planning and design should not be underestimated, there were significant limitations of theory and approach that have left us unresolved problems that are all too apparent in hindsight and as physical legacies.

There can be little doubt as to the need for new approaches. The current debate about "place making" reflects the degree of public and political interest in the problem. Unfortunately the outcome of such debates is rarely a deep enquiry into urban structure and function leading to design innovation. All too often simplistic physical forms emerge that look backwards to an idealised world cleansed of its social and political complexities and absent of modern technology and transport systems. New and sophisticated forms of urban organisation emerge from enquiry and debate and require gestation time. Unfortunately with political timescales becoming ever shorter and framed into media sound bites, this debate often fails to happen.

It is also important to recognise that the city is shaped by power structures as much as by any process of rational design. Circumstances vary from one country to another but in the UK, with its culture of regulation rather than proactive planning, urban form is more likely to be shaped by private landowners than by state intervention. The exception is the provision of infrastructure at both a local and national level. Highway engineering in particular has been one of the few areas of investment in the public realm (arguably since the 1980s), but is one where urban design considerations are subservient to operational criteria, where they exist at all. As a consequence a significant amount of the public investment in our towns and cities has produced poor environments that often have to be renewed before their life span. Poor design is something that we cannot afford in an age of austerity.

There is a considerable gap here between academic writing and research and practice. The nature of cities and the lives of those who inhabit them are inherently messy. Yet hidden beneath this complexity there exist understandable processes at play, which work in different ways to inform the evolution of the urban landscape. In the modern democracies, these then become interpreted into political actions and interventions, seeking to address perceived economic, social and environmental concerns. Without a better understanding of those underlying processes, the outcomes can defy logic or structure and not achieve the aspired results.

Successful practitioners inhabit and understand this world, but often fail adequately to explain it. Their approaches are often based on long experience informed by theory at a very high level, which is then applied intuitively and pragmatically to adapt to the particular circumstances that prevail at the time. But with their practice lacking a strong link to theory, they struggle to argue their case and their solutions when political or economic masters have differing opinions.

It would be a mistake to think that there is a grand theory or algorithm out there that can resolve these issues. And yet by better understanding how the social and economic world interacts with and influences the physical world, the outcomes of urban interventions will forever be only intermittently successful. This book seeks to build on our understanding of how function influences urban form, derived from a fresh look at how we individually and collectively interact with the physical world around us.

Grand theories will always be too coarse-grained to respond to the intricacies of the city. We know most of the principles that make up a human city. We recognise when it is done well. It is about getting the basic elements right and applying them with integrity and tenacity against the forces of expediency. It is about robust narratives and open-ended strategies that are well-grounded in dialogue and research and that are then carefully crafted to suit particular circumstances. Good urban planning and design are not an optional extra. Our towns and cities are too important to be left to chance and narrow vested interests.

Preface

Julian Hart

Travelling from the countryside to the centre of any major city represents a passage through time – from past to present. In that single journey you experience the full history of human civilization.

This is not because the buildings get taller and more sophisticated: the modern country cottage might be made of the most advanced materials known to Man. This is not because there is much more technology in the centre: the modern farmhouse might accommodate all the latest gadgets and be just as well connected into the information networks as any urban flat. This is not simply some aesthetic comment about concreting over the greenness of nature.

Rather, it is an economic observation about the manner in which human beings share space – that is surface land space.

Out in the countryside, land is almost entirely privatized, not shared. Whether it is a farmer's land or rural homes with gardens, each square foot of surface land is treated competitively, exclusive in use to one individual or business. The only shared space is in the form of the narrow lanes and footpaths, which criss-cross the landscape – essential conduits to connect fertile land with human habitations. (It should be noted that national parks and other public features in the rural context are a relatively modern phenomenon decided by and dictated from cities and enacted onto the countryside by urban societies. National parks may sit in and be a part of the countryside, but they are very definitely not a product of rural society.)

Where surface land lies exclusively in the ownership of one or other person (including businesses as legal persons), then human beings are essentially competing for that limited resource. Such competition may normally be quite passive, strictly controlled by a legal framework enforced by the state; but it can be seen to manifest actively as soon as the incumbent owner seeks to sell the land.

As you progress into the centre of human habitations, there is a greater and greater preponderance for shared space. The rural village has its commons. The market town has a library and market place, and probably nowadays a children's play area or two. By the time you reach the centre of a major metropolis, land is extensively shared. For starters, accommodation is stacked; roads are no longer boundary markers, lines of division, but rather represent the streets stitching the urban landscape together, interspersed with public squares; there are parks and numerous public buildings – churches, museums, libraries, hospitals and so on. In fact, in the very centre of the largest cities all surface ground is shared space and much of it is public or partially public space, accessible to all members of society.

The gradual transition from the competition for land in the countryside to cooperation in the use of surface land space as intensity of land use increases, which can be observed on any journey from country into town, is something that has been a product of the progress of civilization.

Appreciating how human competition or cooperation is expressed in regard to the use of space is absolutely fundamental to any comprehension of the urban landscape. It should form the foundation, on which any student of urban planning or urban design should build their knowledge and appreciation of the townscape – past, present and future, rural, suburban and urban.

This book seeks to reconstruct our understanding of urban design around an appreciation of the way that humans experience and interact with each other and with the physical world around them. Out of this, the city comes to be seen as the product of individual action and social interaction; it is form arising from function; it is the outcome of many individual human beings going about their daily lives to satisfy their essential needs to survive and satiate their wants to enjoy another day; it is very much a people-centred approach to understanding the built environment. This way of approaching the city is a far cry from the manner in which the metropolis has been conceived by the mainstream academic and professional worlds of planning and architecture for much of the twentieth century.

For much of the last 100 or so years mainstream urban planning and urban design has been driven by political ideologies: the garden cities movement, the mechanistic city and the liberal city, being just a few examples. None of these have had much basis in a true understanding of the full needs of human beings and how we experience space. In the latter half of the twentieth century and more recently, there have been two principal other influences on urban design theory: these have arisen, on the one hand, from social scientists exploring what went wrong with those previous ideologically driven urban forms and, on the other hand, from architects analyzing what they like about much more historic places, such as old Italian piazzas.

On the latter point, when urban practitioners refer back to those great examples of good urban design from times past, they are often reflecting on cities designed by amazingly intuitive and creative thinkers, who had instinctively perceived how to design places which can best serve the human inhabitants. Alternatively, modern academics may describe those places that have grown organically as examples where form has manifestly followed function. Whilst analysis of the specific physical dimensions of these places can provide a few rules of thumb about what seems to work, it does not, however, educe any theoretical knowledge, which can inform better design of modern cities – especially modern cities which need to accommodate mass private motorized transport.

We need to dig deeper. We need to try to understand how form arises out of function. What is it about the structure of these great places, or any good functioning places, that makes them work so well?

At the beginning of the twenty-first century there is now a sea change in the practice of design of the built environment at different scales. In urban areas design can no longer simply be '*done* to' the resident population, but instead must be '*done with*' local communities. This is forcing design teams to think much more carefully about the real needs and aspirations of those communities. It is requiring architects and their supporting technical teams to be much more person and community centred in their approach. Likewise at the building level, it is becoming increasingly apparent that improvements in energy efficiency and resource use are severely limited unless they are '*done with*' the current or future building inhabitants, working closely with businesses, families and individuals to work out best how to help them exist more carbon-efficiently (cultural changes must accompany the technological improvements). Where many see sustainable design simply as better environmental design and new technology, it is increasingly becoming apparent that genuinely effective sustainable design, and its influence on urban design and building design, is design which focuses around the full set of needs of the human users and occupants (individually and collectively as communities of varying types and sizes).

The outcome of this process is a rejection of ideological thinking, replaced by a more practical and pragmatic approach, desiring to understand how cities, urban landscapes and the built environment generally work from the perspective of those who live in them. This book is a part of that resurgent process and represents an attempt to understand better the whys and wherefores of urban structure, from the macroscopic level of the whole city all the way down to the microscopic, inside individual homes. It will become quickly apparent that there are universalities and themes, which transcend all the scales. Out of this it becomes possible to begin to spell out a *theory of the town* or, indeed, a *theory of place*. Hopefully the insights provided through this work will enable us collectively to be more informed as we plan and design future built environments for people to inhabit.

Wayfinding

For those wishing to dip in and out of this book, rather than read the whole, here are a few comments on the structure.

Chapters 1–3 focus on the structural layout of towns, explaining why town structures have persisted in a recognisable form throughout civilised history, at least until the twentieth century, and then home in on our roads and streets, seeking to explain why and how these differ in function and the impact that has on surrounding town form and urban density.

Chapters 4 and 5 consider the different land uses within the city and how these interrelate (competition and cooperation) across geographic space. This leads to a deduction in both chapters from different perspectives that there exists an urban density threshold above which cooperation of land uses leads to urban agglomeration and below which competition between land uses gives rise to urban dispersal (otherwise known as suburban sprawl).

Chapter 6 seeks to estimate the above noted density threshold and then goes onto discuss the concept of urban density.

Chapters 7 and 8 return to looking at the bigger picture of the city, seeking to explain how urban form naturally changes with increasing density based on the principle that this is informed by the demarcation between the public and private realms of the city. Undesirable consequences arise when the wrong urban form is used for a particular urban density.

Housing is by far the largest built land use across the urban landscape; how it is built has a major influence on both the physical form of the townscape and the quality of lives of a town's inhabitants. Chapter 9 considers the true essence of housing and what this tells us about how it is configured from the scale of the en-suite bathroom all the way up to that of city neighbourhoods.

Chapter 10 brings together all the strands earlier in the book to propose a theory of the town. Every town is different, yet every town is based on an underlying and understandable structural template. It is suggested that there are three major drivers of urban form, which operate antagonistically to each other. The actual structure of a town evolves from a historical narrative of how these urban drivers of structure and form have played out over time.

Chapter 11 seeks to put to bed some misunderstandings around energy consumption of cities and shows what are the most energy efficient urban forms and the physical consequences of pumping high levels of cheap fuel into an urban economy.

Chapters 12 and 13 are focused on the professional audiences, who may wish to understand better how the ideas presented earlier in the book correspond on the one hand to the practices of town planning and property development and on the other hand to the extant guidance on good urban design.

Chapter 14 is then a discursive conclusion, which seeks to ponder on the societal implications of designing our cities in different ways, drawing from the understanding provided throughout the earlier chapters of the book.

Acknowledgements

This work has been a long time in gestation. It represents a distillation into the physical aspects of the city of earlier research carried out during the 1990s on the socio-economic history and evolution of cities, which was sponsored by Ove Arup & Partners and which was undertaken at the University of Surrey. I would like to thank Professor Roland Clift for his support. Many of the original insights that have led to this book came from conversations with colleagues within Arup and in this regard I would like especially to thank Juan Alvaro Alayo, whose observations about the city were particularly insightful. I would also like to thank Lorna Walker, my mentor and line manager at Arup, who gave me the opportunity to spend a semester teaching at the Graduate School of Design at Harvard, which gave me the space and time necessary to explore new ideas and begin the process of developing a genuinely fresh perspective on how the city works.

I would like to extend a special thanks to Neil Deely and his team from Metwork, in particular Jonny McKenna, for contributing all the artwork for this book.

Thanks also to Peter Bishop and John Rhodes for digesting this work and agreeing to provide the forewords.

And finally I would like to thank my wife and family for their patience as I plugged away over many an evening and over many drafts to finally produce something worthy of publication.

Figure 0.1 **'dad at work' by Oliver Hart**

Chapter 1

The Natural Town

Everyone knows how a town or city should be structured. Wherever you look around the world, towns and cities have a familiar layout. Whether Europe, America, Africa or Asia, all towns and cities are generally set out with some shops or a market place in the middle surrounded by homes; larger modern cities naturally express this structure at many different levels and scales from local hubs to international city centre. This structural template, which might be referred to as the natural town (see Figure 1.1), has been the case for as far back into human history as we have records of any form of recognizable civilization. Such structure has seemingly lasted the test of time. Until now.

In the UK at the beginning of the twenty-first century we are seeing our market towns and our suburban town centres dying in their droves, those shops in the middle closing down and moving out of town, leaving boarded up deserted high streets. The old model is seemingly no longer working. Yet, ask any urban designer to design a new garden city or ecotown or suburban extension and their inevitable starting point will still be to think in terms of shops in the middle and homes around.

Figure 1.1 The shape of the natural town

The reason why our market towns and suburban centres are failing is quite easy to establish and will be considered in more detail in Chapter 4. But before worrying about why our market towns are no longer functioning in the way they have for so long, an essential starting point for an exploration into urban planning and design is to understand why such structure has evolved in the first place. Why has such typical layout endured across so many civilizations and geographic regions for so long?

For those unfamiliar with or new to urban design or town planning it may come as some surprise that there is in fact no singular definitive theory that explains why natural towns are structured as they are, as described above. It is one of those little bits of human existence, which is so deeply embedded into our consciousness as being how the world works, that no one has actually bothered to spell out in theoretical terms why towns have shops in the middle surrounded by homes. Why not the other way round? Or, why not have the shops on one side, easy to access from out-of-town, perhaps acting as an interface between townspeople and farmers bringing food to the town?

What is a Town?

Every town is first and foremost an economic entity. By definition a town, from any era and any geography, is a human habitation that has developed beyond agricultural subsistence. It is a settlement where a significant proportion of the population survives in ways other than directly working the land.

In times and areas where most people were or are living off the land, then the natural settlement pattern is and has been that of small villages dispersed across the landscape, providing all members of the population with most immediate access to land to work and grow food. Whilst the agricultural village can support a few people doing a daily routine other than farming (for example, the local blacksmith), once a society has evolved to a point where a sizeable number of people do not toil off the land, then a different form of settlement pattern and structure is required. This becomes layered onto that erstwhile landscape of dispersed villages, where some of the latter grow, by dint of some geographic quirk such as a river ford, larger than the other surrounding villages. Or did trading towns come first, providing a service to wandering tribes, where such nascent cities spun off agricultural settlements to colonise the surrounding landscape? The debate on the origin of cities continues. But, whichever way they formed, the natural town is what it is and is first and foremost an economic entity.

In Jane Jacobs seminal work, *The Economy of Cities* (Jacobs, 1969), she set about showing how each and every town is first and foremost a trading entity both internally and with the world outside. If people are to survive doing something other than growing their own food, then they need a means to satisfy easily their nutritional requirements – each and every day. Most of the time and effort of at least one member of each household must be focused on the activity required to generate an income – acquire money; so, what with all the other chores required to keep a house and raise children, time to source food is necessarily short and should involve minimal effort. The human solution has been the invention of trade, exchanging the product or earnings from some non-food-growing activity for essential nutrition. With the action of exchange itself being potentially very quick and easy (one moment you have inedible money in your hand, then next an apple to eat), the key element of effort is travel to the market place and the transport of the purchased food back to the family home.

So, in any town each household must participate in regular trade to obtain food to eat. For a town to exist and survive, it must as a prerequisite facilitate that such exchange activity can and does happen pretty much every day. If this economic activity were to pause for more than a day or two, a town is immediately in dire straits: there is nothing that can be more essential.[1]

How is a Natural Town Structured?

As noted already, we all know intuitively that every historic town is structured with some form of market place in a central, easily accessible location. But why? The answer is quite straightforward. Locating the market place, or more permanent shops, in the gravitational centre of all the homes is simply *the most energy efficient form*. To appreciate this answer requires further thought on the activity of trade and exchange and how the various different elements of day-to-day human life translate into movement patterns and land uses with identifiable characteristics.

In a time before cars and fridges, as a member of each household in a town would need to make daily visits (certainly most days) to obtain food, especially those more perishable items (meat and dairy), every house needed to be within reasonable access of shops or a market place to buy food. On a day-to-day basis shopping is something which needed (and, to a degree, still needs) to be done very frequently and must not take too much time or effort.

Those shops or market stalls supplying the essential food and materials to the population of a town can be seen to experience many visitations from many of the inhabitants of the town, each passing through quickly for a short period of time to purchase those essentialities before returning home. It can be seen that shops experience a high trip rate and short dwell-times (except, of course, for the shopkeeper himself). In contrast, on a daily basis each house is only generally visited by the occupants of that

1 It is easy to forget this truth within our modern market economies, especially in developed market economies that are remarkably successful at making sure that we all have food to eat on a daily basis. A recent example where the system broke down was when a major earthquake hit southern Chile in 2010. The earthquake caused transport routes to be blocked and so food supplies were interrupted. Within days various towns near the epicentre started to experience food riots. It is quite scary how quickly urban society breaks down, when food ceases to arrive in the market.

household and perhaps the occasional guest; generally each such visitation to a home can be for a considerable part of the day – at the very least for the residents remaining there overnight; the time spent in the dwelling is comparatively very long. Homes, therefore, experience relatively low trip rates with long dwell-times.

For any business seeking to operate in the capacity of a retail outlet, especially for everyday goods, then for it to serve the population as best it can, it needs to locate itself in as accessible location as possible for as many of the townsfolk as possible. The optimal position is the gravitational centre of the town, in a position which gives rise to the shortest journeys and briefest journey times for the maximum number of people. And, if a shop locates in such a position, it will logically generate for itself the maximum amount of trade. In the spirit of Adam Smith's original conceptualisation of market economics, in selfishly seeking to maximise its own trade a shop business manages to serve best the local population by locating itself in the most accessible place for all its potential customers. When those customers are foot-bound, such location is very definitely the gravitational centre.

Describing the economic operation of a town in terms of trips and dwell-times belies a deeper driver. In seeking to minimise time and effort spent sourcing food for the household, people are basically seeking to minimise energy expenditure. *The logical conclusion of this train of thought is that the actual and most fundamental driver behind the structure of the natural town is energy efficiency.*

Before humankind had discovered black gold (oil), energy efficiency was paramount.

Various other models of urban structure were developed over the twentieth century, including those of Burgess (Burgess, 1924), Hoyt (Hoyt, 1939) and Harris and Ullman (Harris and Ullman, 1945). None of these models contradict the argument posed above and below. These earlier models were largely formulated to explain the more detailed urban structures of larger cities, such as Chicago. None of these models formally rely on the notions of trips and dwell-times, nor do they make any reference to energy efficiency; they are, however, all predicated on the notion that the centres of larger cities are necessarily and inherently the most accessible zones and also the areas of highest land value. A fuller list of extant models can be found in Wikipedia under *Urban Structure*.

Looking elsewhere in the academic world, one would have thought that this thinking might have been visited before in the field of transport geography. But that discipline appears to be premised on the notion that urban structure is inherited, largely a legacy of the past, and that the role and purpose of the field of transport geography is to consider how to connect existing destinations in the best way possible (i.e. optimising the efficacy of the transport infrastructure).

The closest theoretical model to this way of thinking (natural town structure derived from an energy efficiency driver) is that of Central Place Theory (a complete explanation can be found in Wikipedia under *Central Place Theory*). Within the assumptions underpinning Central Place Theory there is a factor for transport costs, with transport cost being directly proportional to distance. In this respect energy and effort of travel is implicit in the model; it is assumed that travel costs will be minimised by individual agents. Central Place Theory was originally intended to describe the pattern of human settlements across the geographical landscape, rather than the structure of towns themselves. However, when it is extended to a situation where all the landscape has become urban, then it quite apparently gives rise to a multi-nodal version of the natural town (high density centres, surrounded by a lower density suburban landscape) and has been used to inform the planning of larger cities, such as London. (More consideration will be given on the structural form of large cities in Chapter 10.)

The key message to hold in mind is that the form of the natural town, the town we love, hold dear and with which we are all familiar in terms of its general structural form, is and has always been determined by the need to minimise effort by its inhabitant population in order to function optimally. That is why we have (or have inherited from our existing towns) a typical structure where the shops are in the middle. While ideas such as Central Place Theory effectively end up with the same result, no theories of the town to-date formally recognise the fundamental underlying driver of historic urban structure – *energy efficiency*.

Land Uses

Using the concepts of trips and dwell-times, introduced above, it is apparent that houses and shops lie at two ends of a spectrum. This is depicted in Figure 1.2.

Within this spectrum there exist other land uses. The most prevalent other land uses correspond to buildings or locations used for business purposes other than retail, specifically offices, small-scale warehousing and high value manufacturing;[2] these other land uses do not depend on such high daily visitation levels; only those working in the business and a relatively few other people, who have dealings with the business, need to travel to its premises; dwell-times for visitors (working day or attending meetings) will be on average longer than for individual shops and shorter than for homes. Therefore, business premises, including warehouses, lie somewhere in the middle of the spectrum between homes and shops; it is natural for them to be located in and around the town centre, not quite at the most accessible locations, but near to.

The three major elements of any town (housing, businesses and shops) lie on the main spectrum: the number of visitations that these types of land uses experience on a day-to-day basis is (at a statistical level) quite predictable. There exists another axis to the diagram above, which correlates to variability in visitations experienced by a particular land use (see Figure 1.3). A home or shop

2 Remembering that this is the natural town being considered, where our modern urban landscape over the last 50 years has strayed far from this long-standing structural form. And remembering also that the heavy industries of the Victorian age were, for that period, high value manufacturing.

will typically have the same consistent level of visitations every day of the week; a sports stadium in contrast may experience many visits on one day and very few for the remaining six days of the week; a hospital may experience significant variation in visitations over a different cycle – perhaps more in winter-time and less in the summer.

Figure 1.2 Diagram of trips and dwell-times on two axes

Figure 1.3 Diagram of trips and dwell-times on three axes

These other types of destinations will also likely experience higher levels of variability in terms of who is actually visiting them: for example, those who may be visiting a hospital in one quarter of a year may be very different to those who may be visiting it in the same quarter of the next year, whereas a group of local shops will see a strong consistency in who is visiting them day-after-day, month-after-month and year-after-year. These influences driving the locations of these other types of premises are thereby subtly different to those influencing the location of retail, other commercial and housing land uses. The latter are strongly dictated by accessibility (minimising journey times and effort); the more variable land uses are not influenced by these factors to quite the same extent. The optimal location of these other types of land uses are, therefore, more dependent on local circumstances and historical narrative (how the town grew over time) than forced by accessibility.

Where a location is being used by a business or for some economic purpose other than living, then the land is being used to generate an income. But businesses do not need to be in the prime accessible locations, like retail, so they can avoid paying the high rents associated with highest land values at the very centre of a town on the most accessible street in the town – the high street. However, in using land to generate an income businesses can afford to trump (at least a proportion of) higher value homes for the next best accessible locations. The outcome is the typical town structure with shops in the middle, businesses in slightly less accessible locations in and around the centre and homes surrounding or sitting above this commercial heart, with other land use types being dotted around the town in a more random pattern. There are always exceptions; but this is the general configuration that we recognise.

Land Value

In a world or time where a majority of the human population of a town were foot-bound, then accessibility gave directly to land value (further discussion on exceptions to this generalisation is provided below). More accessible locations, which by definition are more centrally located places, automatically became higher in value. Rather obviously, a person living in the centre of town would have to spend less time and effort walking to the market to purchase essential foodstuffs and other materials than other inhabitants living further out. In addition to this, plots of land in the centre of a town represent locations where it is possible either to establish a retail premises selling goods to the town's population or is better positioned for businesses to operate.

Moving away from the centre of town, however, land becomes almost entirely dedicated to human habitation – homes (the degree to which housing is the dominant land use across the townscape will be explored in Chapter 5). In not being so suitable for retail or business use, land away from the centre has a value directly proportional to the degree of accessibility to the town centre (as a proxy measure). There may always have been exceptions, but, as a general rule, in the historic natural town the more effort (both time and energy) that had to be expended to reach the town centre to obtain daily goods the lower the value of the location.

The structure of the typical town is analogous to the land use patterns envisaged by Von Thunen, an early academic in urban geography. He developed a concentric ring theory (concentric circles around a town), which explained the use of agricultural land as you moved away from a town or city. In Von Thunen's model, the drivers behind location of different types of agricultural production included value of the goods (often dictated by degree of portability and time it took to 'go off'), distance and weight of goods (informing level of effort in transport) and land area required to make the produce. The output of his model typically put vegetable, fruit and dairy production close to the town, woods further out (heavy to transport) and then meat production (easy to transport – if you make it walk itself) and arable farming (requires large amounts of space) furthest out (see Figure 1.4). Von Thunen's model, as originally conceived, strictly related to the surrounding countryside and not the structure of the town itself.

The driver behind the concentric rings of different uses and thence values of land within a natural town is, however, fundamentally the same – ultimately dictated by energy consumption in transport. For the different land uses, the various trip and dwell-time patterns have then to be layered onto this underlying energy driver. For the housing, a key difference arises from each household needing to obtain the full gambit of goods on a regular basis – all types of food, wood, etc.; so, criteria such

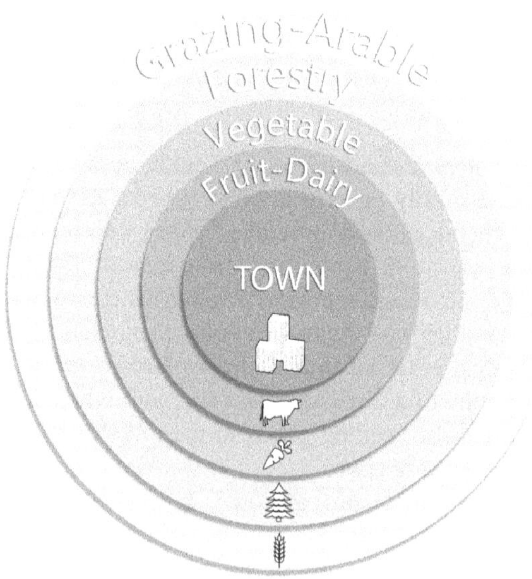

Figure 1.4 Depiction of Von Thunen Model

as transportability of individual goods becomes irrelevant; instead the singular factor is time taken and level of effort to reach and return from the town centre, the source of all goods. For the housing, then, instead of there being distinct concentric rings, there is a gradual reduction in land value with reducing accessibility.

In a time before the automobile, the accessibility map of any town and the land value map of a town were essentially commutable – land value directly correlated to accessibility, mediated by the corresponding effort in travel and transport. This is a generalisation, but broadly holds. There were, no doubt, exceptions to this rule in each and every city, caused by, for example, continuing social structures and hierarchies and cultural factors: for instance, wherever the King chose to live, the immediate surrounds became higher value. These social factors, which distorted the underlying economics, were more pronounced in earlier times and for smaller towns; but as the human population has grown around the world over the last millennia and towns have expanded in size, the macroscopic maps of all towns, anywhere in the world, have converged towards this principle of being structured according to most energy efficient form, with the market place located at the centre. Prior social or cultural determinants have step-by-step given way to economic influences. Before the motor car, such structure was stable and durable – energy efficiency dictated accessibility, dictated land value.

Bring on the Car

The materialisation of mass, motorised, private transport has thrown this historic pattern of development into complete chaos (the word *chaos* is used quite literally here, as will be seen later in this book). Where those derived land use patterns were fundamentally formed as a result of the effort and time taken for travelling by (mostly) foot, the horseless-carriage has changed everything. Travel distances have been effectively shrunk, by a very significant degree. In your average historic market town (or suburban centre) it now takes no effort and essentially next to no time to access any part of the town. What was a 10 to 15 or more minute walk is now less than a two-minute ride. Furthermore, with a car (and a fridge at home), one can easily carry home a whole week's worth of groceries. Before the car, the idea of carrying more than two days' food supply back to the home would have been seriously onerous. But now … people no longer need to food shop every day. This new technology (car and fridge combined) has changed the trip characteristics of the food aspect of the market place, which for much of human history has been the major part of the heart of any town.

The outcome (erosion of distance and changed trip/dwell-time characteristics) is that the food element of the market place no longer has to be located in the gravitational centre of the town. The food shopping can be located anywhere, outside the town or even on the far side of the next town (or suburban centre), and still be easily accessible to anyone with a car. Previously unimaginable travel distances have collapsed to take little time and zero experienced effort. Of course, we all know that the actual energy consumption is vastly more than before; but in the context of relatively cheap petrol and diesel and that driving a car takes little effort, the real level of energy consumed is irrelevant. It is what we can each perceive and afford that counts and, critically, time spent; probably time spent more than anything else; issues such as time wasted finding a car parking place have become more important than overall travel distance.

The impact of this on our market towns has been understandably severe. In retrospect it has been highly predictable. The centre of accessibility, or more pointedly the location of maximum accessibility, is no longer the gravitational centre of the existing housing stock. Rather it is dictated by the width and fast-flowingness of roads through and around the town. It may well be, as our supermarket chains were quick to discover, that the easiest and quickest place to get to by car might be situated on a new bypass around the town or halfway between two towns. Combine this with the ease of car parking in newly constructed out of town facilities to minimise time taken on the journey, and the decay, and sometimes complete dissipation, of the historic town centre high street has become an inevitability. The car, with all the freedom, liberation and benefits that it conveys to each and every one of us, has a lot to answer for.

Whilst many of us may rue the loss of our market town centres, does it really matter? Clearly at present global warming is deemed to be the principal problem associated with all of us using our cars. But imagine if we were able to develop technology so as to have carbon-free private transport, that emitted no air pollution and created no noise. Does it still matter?

As a matter of fact, it does indeed matter. It matters very much. The car changes our whole urban landscape in ways that are obvious – noise, air pollution, energy consumption – and in other ways that are not so obvious, but which have a dramatic effect on our whole societies. The impact that the car has had on our towns and cities does, however, provide a means to understand better how our towns naturally function, and how function gives rise to form. This in turn provides some of the insights necessary to begin to build a better understanding and knowledge base on which to build our disciplines of urban planning and urban design. The place to start this journey of intellectual exploration is on the *road*.

Chapter 2
A Journey into Town – Part 1

Historically, the overall structure of a town arose from the members of the inhabitant population all seeking to satisfy their basic economic needs in as energy efficient manner as possible. Stripping away the luxuries of town life, as a bare minimum people need and needed homes to live in, places to go to work to earn an income and shops or a market from which to acquire essential food and materials to survive. A stable structure emerged, herewith referred to as *the natural town*, which persisted for thousands of years.

In the very earliest cities, such as Catal Huyuk (Anatolia/Turkey) some ten thousand years ago, things were a little different. Inhabitants seemed to have combined all the essential activities together in one abode, such that every *hole in the ground* represented home (including burial chamber), work place and shop all at once (where it has been deduced that the shop aspect enabled inhabitants of the town to exchange products manufactured in the home for food brought into the town by visitors). The whole town was the market place; to access each dwelling/work-shop, other members of the town or visitors would presumably have had to walk over or along the walls of other surrounding homes; the archaeological investigations suggest that there were no roads, as such (Çatalhöyük Research Project, 2013). As cities have evolved and grown beyond what was (by today's standards) a village, this solution proved quickly to be impractical; at the very least arterial roads became necessary to allow access from the countryside into the heart of the town. This evolution of the town structure operated hand-in-hand with an increasingly apparent division of labour between those doing the buying/selling (trading) of stuff and those doing the making of stuff, where the traders took to operating from a singular location, the marketplace; from very early in human civilization, the marketplace became a discrete and normally central entity within the town, accessed by those arterial roads.[1] Ever since, roads have been an essential feature of the townscape, not just the rural landscape. But, as anyone who has spent time living in a busy city will know, necessary as roads may be, the traffic along them can be both a nuisance to and divisive of the communities through which they pass.

Within the modern professional worlds of town planning and urban design one of the most apparent and recurrent design tensions to be resolved in any masterplanning exercise is how to enable access to and through parts of a town while simultaneously creating comfortable places for people to live. In the professional world of urban planning and city design, this has become expressed

1 Further reading in support of this chronology and explanation can be found in Kotkin (2005), Mumford (1961) and Reader (2005).

as a fault line between the traffic engineering fraternities, seeking to maximise speed and flow of traffic on the road network, and those design professionals hoping to promote placemaking, more often than not perceived to represent the creation of pedestrian friendly urban environments.

The design objective to make the road network more effective and efficient clearly arises from an economic imperative to get food and goods into and dispersed throughout the city, to dispose of waste generated by the population and to enable export of any manufactured goods. With the dramatic growth of cities during the twentieth century, over the last 50 years a whole intellectual industry has evolved dedicated to the cause of engineering faster roads with greater capacity; it has emerged as a hard science based on rigorous mathematical algorithms incorporated into sophisticated computer models; it is entirely predicated on the premise that the ideal transport network minimises all travel journey times for all.

In contrast *placemaking* (as it has become known) remains an elusive art. It is difficult to pin it down to any specific objective other than, perhaps, aiming to promote pedestrian activity. It is more easily defined in terms of what it is not, as a counter-play to fast flowing roads and lots of traffic, than what it is. For so long as this aspiration to create *places* remains so ill-defined, it will continue to be a hit and miss venture, with success rarely, or at best only accidentally, sustained. Without providing a sufficiently strong opposing tension to the traffic engineering objectives, then cars will increasingly dominate in all but the most central urban locations

For new theoretical thinking in urban design to have any success in improving how we design our towns and cities, it needs to make more explicit and manageable this ubiquitous design tension between roads and places. A first step forwards is to formulate a formal definition for *roads* and *streets* and how these two elements of the townscape differ from each other. Counter-intuitively, to begin a study of roads and streets, the best starting place is not in town, but out in the countryside, where roads really are, more or less, just roads. Then we can travel into town to explore how *roads* morph into *streets* and how these elements of the townscape differ from each other.

The Road in the Countryside

Outside towns the primary purpose of any road is to connect destinations to enable communication and the transport of goods between human settlements or between any human settlement and its food growing hinterland. In that they cost money to build and maintain, without such economic raison d'etre roads would simply not exist; their primary role is for trade and communication; they act as channels of goods, people and information to allow economic activity to flourish; they provide the shortest, quickest and most energy efficient means to connect destinations. Furthermore, at this functional level, road and rail can be treated interchangeably.

As already asserted in Chapter 1, in the natural historic town, anything from the Mesopotamian civilization forwards to last century, most of the economic activity within a town is necessarily channelled through its centre, its core. So, any road fulfilling an economic function leading to a town is naturally going to go straight to the heart of the town to serve the market place. These arterial roads into a town consequently become the largest roads running through the urban landscape and give rise to the greatest challenges in terms of the design tensions to be resolved, as noted above. More consideration on how to go about resolution of these design tensions will be given in Chapter 10.

Some of the following statements regarding *roads* may superficially seem overly obvious, even trite. But they are actually essential analysis to support what follows with regard to understanding the true character of *streets*.

As a connector between two locations, the road is a linear object running through the geographic landscape. What happens on either side of the road is, to all intents and purposes, irrelevant to its primary purpose. The adjoining landscape may provide passing interest to the passenger in a vehicle – *passing* being the operative word, a fleeting view and then gone, passed by. For the traveller along a road, their focus and purpose for being on that road is in general to travel along it, towards a destination on the highway, as fast as safely possible.

Think about how you experience space, when you are travelling along a road, any road, and in particular when you are in a car. On the road you have a singular direction of travel along the length of the highway towards an unseen destination (unseen for most of your journey); the modern highway is further divided into lanes, in order to minimise the impact of traffic running in the opposite direction. The highway itself might bend (as a hosepipe channelling water might bind), but your direction of travel unerringly follows the line of the road, within your lane of the road; you are enclosed on either side with only a fleeting (if any) interest in what you are enclosed by; behind you is your past, where you have been and from where you have left; you are essentially in a time tunnel, waiting to get to your destination, on a single dimensional journey to the future. The faster you travel, the more the above observations hold true.

Now consider the way a road interacts with the landscape through which it runs. For a better understanding of urban design, this is much more than a passing interest. For, whilst a road through the rural landscape seeks to connect end-to-end, it simultaneously acts to divide side-to-side. This is most obvious with respect to the modern multi-lane motorway.

The wide motorway dedicated to trucks and cars represents an impenetrable line, cutting through the geography of a region, clearly dividing the land into that on the right and that on the left (see Figure 2.1). It becomes a physical rift; where once two villages may have been a walking distance apart, they become potentially separated by many multiples of the crow's flight; where communication may once have thrived between those two villages, in the absence of a direct bridge over the motorway the communities will physically interact no better than the actual new longer (potentially contorted) travel distance, which has to be negotiated to get from one to the other.

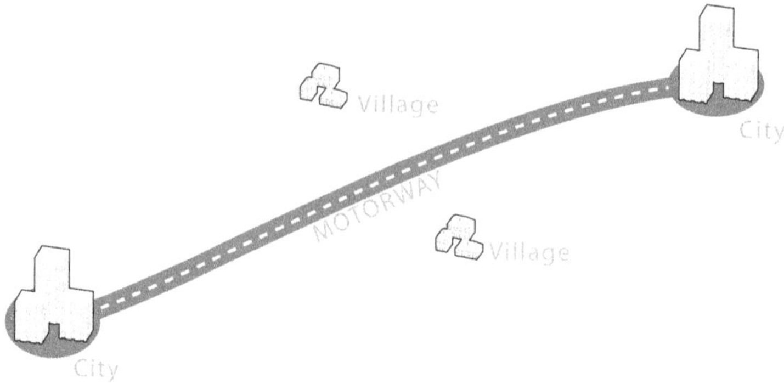

Figure 2.1 Road connecting at the ends and dividing to the sides

This observation of the way motorways act to divide the landscape is not meant to be pejorative; it is a fact (a non-negotiable impact); it is a consequential effect of the purpose of the highway. All the time the towns at either end of the motorway benefit through increased communication and trade, the landscape either side is inevitably divided; the wider the motorway and the faster the traffic along it, reducing travel time and increasing interaction and economic activity between the ends, the more it divides the landscape through which it runs. From the perspective of a person living in a village close to the side of a motorway, their *Accessible Space* (see Box 2.1) is quite similar to that experienced by seaside or riverside villagers, albeit with none of the amenity, beaches, sea or river views to compensate for the loss of accessible landscape. In these contexts, the village inhabitants can access 180 degrees of land around their habitation, instead of the normal 360 degrees.

This blatant inequitableness between those who benefit from and those who are disadvantage by new motorways (or rail routes, for that matter) is what gives rise to the social angst and political arguments that such projects inevitably provoke.

Box 2.1: Introducing Different Types of Space

In our daily lives as we go about the country or city, we experience different types of space around us. Perhaps the most important for the purposes of urban design is the notion of Accessible Space. This is the space on the landscape to which you as an individual have access from wherever you happen to be. As a stranger in the city, this is the public open spaces – the streets, squares and parks and public buildings. In the countryside, depending on the rules of the land, this will be just roads and footpaths. In the wilderness or a national park, the Accessible Space will be dictated by the geography and topography of the landscape – a river or cliff may create an impenetrable barrier – otherwise the whole landscape is accessible.

Another type of space is Visible Space. Standing in your own garden, your Accessible Space is clearly demarcated by the fence or hedge around the garden. But your Visible Space will extend much further, across all the other neighbouring gardens. Within buildings Accessible and Visible Spaces tend to be far more coincident. But turn the lights out and Accessible Space may become larger than Visible Space.

Definitions:

Real Space is the real space around an individual, ignoring any limitations to accessibility.

Accessible Space is the space that an individual can access from their current position, according to legal, social or physical norms and constraints.

Visible Space is the space that can be seen from a current position.

Audible Space, Olfactic Space, Tangible Space and Gustative Space each relate to the other senses. These are probably only relevant in the context of specific design problems, such as designing an auditorium or for the blind or deaf or the creation of an emotional connection to a space.

Experienced Space is a combination of Accessible Space and all other spaces related to the senses.

The Road in Town

As you journey into town on a major thoroughfare, the impact of an urban motorway changes little to the way it affects the rural landscape. The major arterial roads provide the means to bring people and essential goods to the town's centre, its market places and business centres. The arterial road is also the quickest way to take goods, purchased or manufactured, and waste products back out of the town again. The collective activity of the town's inhabitants, feeding the population, removing wastes and exporting products, especially the successful economic functioning of the market place in the town centre, are totally dependent on the unrestrained teeming of humanity and goods along the town's major arteries and veins. The road into/out of town must be as direct and unobstructed as physically possible.

Urban motorways slice through urban communities, splitting them apart. It is this impact of motorways in urban and suburban areas that so vexed Jane Jacobs to drive her to write her first book, *The Life and Death of Great American Cities* (Jacobs, 1961), a major theme of which was the disastrous effect on New York's communities arising from new (1950s and 1960s) motorways being built through historic urban landscapes.

Out in the countryside, when a new motorway (or railway) is planned, the rural village has little option but to accept its fate, to inhabit a new *Accessible Space* of half the size, cut off from previous neighbours. Crossings are intermittently put in place across motorways; however these are usually part of junctions created to connect larger towns on either side and to enable fast access to those larger destinations. Villages to east and west or north and south are simply divided, split apart for good; they do not warrant that level of investment required to bridge the newly created gap in the landscape.

Within cities, given the closer proximity and higher density of population and economic activity, the impact of the division generated by an arterial urban motorway is more keenly felt by society. Furthermore the increase in travel distance between erstwhile neighbours may be many multiples of the previous distance and travel time: not just double or triple. To address this, various solutions are regularly used, including burying motorways or train-lines in tunnels, or creating undercuts (which make regular bridges easier and cheaper to construct), or raising the motorways or railways up on stilts (consider the M4 and M40 approaches into London or the metro system in Chicago). These all represent significant infrastructure investments, costing significantly more than the civil engineering required for a ground level road or rail track.

In her incisive book on American cities, Jacobs was riling against the extreme impact of major motorways on once cohesive urban communities. Community studies indicate, however, that a road does not need to be a major motorway for it to act divisively on town communities. Work originally undertaken by Appleyard (Appleyard, 1969; 1980), and more recently repeated and up-dated (Hart and Parkhurst, 2011), demonstrates how the more traffic there is on a road, even relatively small roads, the more it will act to separate communities living on either side. In the above referenced paper, Hart concludes:

The results confirmed that Appleyard's findings are applicable to the UK in the twenty-first century; specifically that the number of friends and acquaintances reported by residents was significantly lower on streets with higher volumes of motor traffic. The extent of people's 'home territories' also diminished as motor traffic increased. Other notable outcomes from the research include the finding that individuals' perceptions of road safety in their neighbourhood may be disproportionately influenced by the traffic conditions on their street of residence, especially affecting the degree of independence granted to children.

Yet not all carriageways separate the landscape on either side. A little logical reasoning, supported by the output of the above noted research, suggests that not all highways will be so disruptive to the abutting urban or rural landscape. The short residential cul-de-sac, serving a group of homes, is not likely to be divisive; rather it will act as a piece of public realm, enabling interaction between the inhabitants of the houses around and along the cul-de-sac; it represents an environment where local children can play and interact without fear of being mown down by a horseless-carriage. And a high street, especially one with a strong pedestrian element, is not logically going to divide its urban surroundings, but rather help to connect the two sides, the two lines of shops.

If a *street* does not function like a *road*, what, then, is a *street*?

The Street

By the time the visitor to a town has reached the very centre of the city, the roads have taken on a very different character. At the opposite extreme to the multi-lane motorway, there exists the shopping street or the alleyway through the souk. On such shopping street the raison d'etre of the shopper to be on that roadway is no longer to get from one end to the other as fast as possible. Instead, the major interest has completely switched to that which lies on either side of the street, not what lies at either end. The shopper strolls along, peering into and nipping into the shops lining both sides of the urban corridor; the end of the street is largely irrelevant; in the shopping street the focus is wholly on the sides. The typical, well-functioning high street may still be a linear object, still appearing as a line on a map, perhaps still constructed out of tarmac, but the way that it functions and the role that it plays is fundamentally different. It is not a road.

A major and rather obvious design difference between a road and a true street is that streets have no lanes; traffic on a street can move in any direction on any part of the street, whereas on a road, the direction of travel of traffic is strictly controlled to defined directions.

To understand the full degree to which streets differ from roads it is instructive to analyse the thinking behind the interior design of modern shopping centres. Out-of-town shopping centres, or Malls,[2] may themselves be the antithesis of what is perceived to be good urban design (for reasons, which will become very apparent in later chapters); but within them they contain malls, which are designed to be perfect high streets. Appreciating the logic behind how such malls are constructed helps to explain how the ideal street functions.

Retail centres are nowadays structured according to a series of rules of thumb, which are known to work – not for any aesthetic reason, but for the sake of brute economics. Some very specific design measures have been found to maximise the income for the retail tenants and thence the landlord. One of the most important guidelines is that the layout of the retail centre must have anchor stores (in the UK these are normally M&S, John Lewis, Selfridges, Debenhams, House of Fraser and the like) at the ends of the malls. All the remaining shops line the street or streets running between the anchor stores. Typical layouts arise such as the dumb-bell, the triangle or the cruciform (see Figure 2.2).

2 Through this book, Mall with a capital 'M' will be used to indicate whole shopping centres, whereas mall with a small 'm' will be used to refer to the street within a Mall.

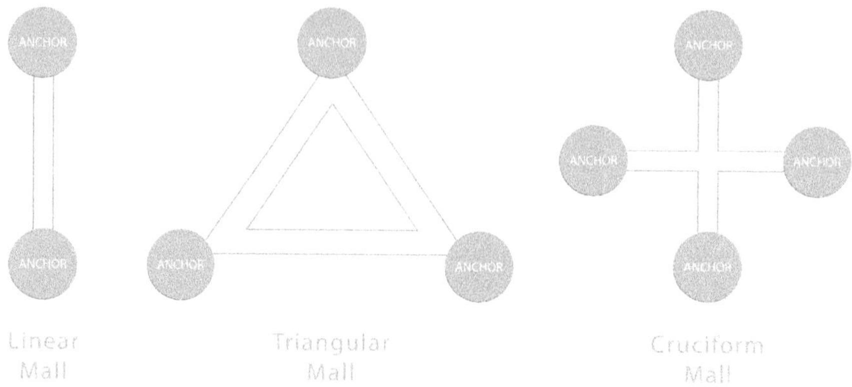

Figure 2.2 Typical retail layouts

For the retail developer the logic behind these structures is that the majority of the landlord's income arises from the rent paid by the shops lining the streets. The anchor stores represent *draws* to bring people to the retail centre as a whole; in return such stores pay a significantly reduced rent. To give a sense of the scale of this, a typical anchor store may pay, say, £10 per square foot (albeit on a large area), whereas a shop along the high street of the same retail centre could pay upwards of £300 per square foot for Zone A (the front 20 feet of their shop), halving to £150 per square foot for Zone B (the next 20 feet) and so on; even by the back of such shops, the tenants will be paying, say, five times the rate of the anchor stores. These are big differences.

In charging such high rents, it is critical for the Mall developer that his retail tenants will be successful.[3] To seek to ensure such success the layout must aim to maximise the footfall (the number of people walking past the shop front per unit of time) experienced by those shops along the streets in the Mall. However, footfall alone is not sufficient. The shoppers need both to *amble* and to *meander*; the majority of the visitors to the shopping centre should ideally walk slowly past the shops maximising both the number of times that they pause at shop windows on either side, regularly entering shops, and the degree to which they traverse from side to side, as they wander (see Figure 2.3).

3 In modern retail centres rents for shops are often determined in part on basis of a percentage of the retailer's revenue. Landlord income is thereby dependent on success of his tenants.

Figure 2.3 Meandering shoppers

Here we see one of the most obvious physical differences between roads and streets. As already discussed, the sides of roads are, to all intents and purposes, impermeable, except where they cross other roads. Furthermore, not only does the pure road represent an impenetrable rift through the landscape, in being a single dimensional linear object it is entirely self-contained, keeping those on the road focussed on driving (lorries, cars, horses or cattle) towards their destination, ideally within their lanes; and keeping everything else, errant deer, kangaroos or other, off the road (as far as is possible and practical). Large roads and motorways are often hoarded along their sides, especially through urban areas, to reduce noise; this further accentuates the isolation of the road from the surrounding landscape.

In contrast the ideal street is highly permeable; it is leaky; it has a sieve like membrane along its edges, enticing those who are on the street to interact with its sides, to leave the public realm and make frequent sorties into the territory on either side. In having permeable edges, the street itself represents a highly permeable object through the urban realm. If well-designed it too entices those off the street onto it, offering opportunities to explore and experience all those other shops or market stalls that line its sides.

The retail developer's objective is to make each mall within their Mall exhibit the characteristics of the ultimate street (the absolute opposite to a road). To achieve this effect, he designs the shopping centre just so; the malls must not be too long; they also need to be just the right width; and, intriguingly, it is essential that the anchor stores at either end are visible at all times for those on the street.

Now, from the perspective of someone new to retail design, when first exposed to these retail development rules of thumb, the condition that says 'the end anchor stores should be visible at all times to the people on the street in between' makes no obvious sense. The vast majority, probably over 95 or more per cent, of all shoppers in any shopping centre at any one time are people on return visits (at least once it has been open for a few weeks or more). Surely they know their way around? Why do they continue to need to see the signs for the anchor stores at either end of the streets?

The answer to these questions comes down to the psychological effect for human beings of having a clearly bounded space. If the shoppers were unable to see the anchor store signs at either end of each mall, then they would be influenced to act more as if they were on a road, than a street. As already noted, on most roads for a majority of any journey you cannot see your destination (certainly not in the modern car-borne world). If shoppers cannot see the ends of the mall, the retail architects know that they will have a greater tendency to walk along the thoroughfare taking less notice of that which lies on either side. By clearly bounding the shoppers with very visible ends to the street, the retail developer influences a statistically significant number of shoppers to slow down their pace and focus more on the sides of the streets than striding out to reach the unseen ends. It is subtle. We may each rile against the notion that the shape of the place that we are in might influence our behaviour. But it does. Much of the modern retail industry is dependent upon it.

The pedestrian street within the shopping centre has taken the concept of the *street* to its logical conclusion. The ends are, in fact, now competitive entities; the major anchor stores all sell pretty much the same stuff; they are in direct competition, both physically and economically. From the perspective of each of the anchor retailers, they do not actually *want* any shopper ever to reach the other end of the street; if it is well-designed and provides sufficient interest along the route, shoppers will tire in their meandering and return before they ever reach the opposing anchor store. In contrast, the shops on either side are reliant on maximum connectivity across the width; they may be competing economically, but (as will be discussed in more depth in the following two chapters) these smaller retailers are actually cooperating in physical terms. They are creating a seamless, permeable edge with continuous interest for shoppers to explore as they make their way, slowly, along the street.

Placemaking

Drawing all these observations together, it can be concluded that the ideal *street* functions orthogonally to the pure *road*. Where the role and purpose of any road is to connect end-to-end, the ideal street is designed to be so full of human bustling activity that no one ever reaches (or wants to reach) the ends. Where the road is an impermeable, single dimensional object running through the landscape, the street is a two-dimensional[4] space with highly permeable sides. Where the road divides the landscape on either side, the street connects and stitches the physical worlds on either side together. Where the success of a road might be measured by the quantum of traffic travelling along its length, the success of a street is defined by the amount of human interaction across its width.

4 The dimensions of different spaces in the town are explored further in Chapter 3.

This latter point is worth exploring a little further. The quantity of traffic on a road is clearly defined by the degree to which it connects end-to-end, enabling economic activity to flow between the destinations at either end – the scale of commerce between two connected towns. Consider in contrast the case of a passageway through a souk, an ideal street, where stalls line either side and transactions are taking place over the stall tables, essentially happening across the line of the edge of the street. The economic success of such street is defined by the quantity of economic activity (money and goods exchanging hands between shoppers and stall-holders) taking place across the sides of the passageway, across the sides of the street.

It can now, hopefully, be appreciated how, in economic terms, the *road* and *street* are functioning orthogonally to each other. It is no wonder, then, that where congestion on a road is perceived as a failing, hindering its prime function, congestion on a street is simply busy-ness and is a mark of success.

There is presumably a spectrum running from one extreme to the other (multi-lane motorway to busy high street), where many of the roads/streets that we experience on a day-to-day basis, outside our front door, along which we travel to take the kids to school, to get to work and where we might shop, lie somewhere along this spectrum. It may be quite subtle: the degree to which any urban road might act as a tear or a seam. It may require scientific studies, akin to those done by Appleyard and Hart, to enable observation of the differences. But imperceptible as the effects may be from a quick visual inspection, they exist and they impact on the way our towns and cities function, the quality of life that we each experience on a day-to-day basis and the degree to which our communities are nudged together or torn apart by the traffic flowing through them. If we design our urban areas with no appreciation of the differences in impact between roads and streets, then all talk of creating cohesive communities will likely be in vain.

Roads and streets may both be lines on the map and both be lain with tarmacadam. But analysing the function that these two elements of the townscape play, it is quite apparent that they have orthogonal purposes: hence, for example, why roads are generally divided up into lanes in order to facilitate that longitudinal motion, whereas the ideal street has no lanes and allows movement in all directions on any part of it. More consideration of the consequences of this realisation will be given in the next chapter. At least, however, by formally defining *roads* and *streets* as having perpendicular objectives, it begins to explain and allow us to begin to better manage the inherent and probably unavoidable on-going design tension between those traffic engineers and the placemakers.

Chapter 3
Making Places and anti-Places

The function of roads in human society is quite obvious: for our economies to work, people need to move themselves and their goods between relatively distant destinations A to B and back again. In order to operate in this capacity, roads are inherently transient places; their whole intent is to facilitate movement – the faster, the better. As set out in the last chapter, the role played by streets is equally important in a very different way, but being linear objects on the map streets have long been confused with roads; they do often, after all, look very similar; but streets are not roads; streets are not simply roads with less motorised traffic and a greater proportion of pedestrians; they have a fundamentally different property. In the context of the high street or mall, the role of the street is to maximise local economic interaction, to facilitate the action of exchange in one specific area of the town. Roads, in contrast, enable trade between distant locations. The irony, here, is that in conceptual terms a road always has to terminate in a street (of some form); the road facilitates the movement of people and goods, but it does not itself enable the subsequent essential economic transactions to allow trade to flourish. In other contexts, as will be seen, the street functions to enable social interactions, instead of economic ones.

Failing to realise that the function of a street is to facilitate lateral movement in the urban landscape can be seen to be, and have been, rather a handicap for the practice of urban planning and design. Once the real difference between roads and streets begins to be appreciated, it gives rise to a whole new understanding about how our townscapes operate. This will be explored over this and the next chapters.

Dimensions of Space

Through the process of thinking about roads and streets in terms of both the function they play for human society and the way that people behave and think when on them, it begins to be apparent that there exists a wide spectrum of different highways;[1] not all strips of tarmac are the same. Pure roads represent one-dimensional objects weaving through the landscape, whereas streets create two-dimensional public realm. For those not familiar with the concept of fractal dimensions, this requires a little further explanation.

---

1 Now appreciating better how streets function, the term *highway* doesn't feel quite so appropriate as a collective noun for roads and streets. What should be use instead?

Fractals are systems, which can be explained in precise mathematical terms. Many natural systems are fractal by nature – trees, snowflakes, your lungs and capillary system, coast lines, the surface of a cloud, mountain ranges and numerous other objects with which we are very familiar, including cities. Benoit Mandelbrot was the genius who discovered the mathematics behind fractals. He came to realise that he could describe these natural systems through a concept known as fractal dimensions. Most of us are familiar from school mathematics with the Euclidean orthogonal dimensions – north and south, east and west, and up and down – making three perpendicular dimensions. Mandelbrot discovered that the surface of a cloud could, for example, be described in mathematical terms as something that has two and a bit dimensions; it is more than two dimensions, because it is not flat; but it does not fully occupy three dimensions. The structure of trees also displays two and a bit dimensions: trees do not fully occupy three-dimensional space, in the same way that a solid object might. A tortuous coastline, and for that matter a winding road, can be defined mathematically as being somewhere between one and two dimensions: the straighter such object is, the closer it will be to a single dimensional object; the more wiggly, the closer it will be two a two-dimensional object.

More recently Michael Batty from University College London has used the concept of fractals to describe city structures (Batty and Longley, 1994), in particular the manner in which cities occupy surface land. He has shown that cities have fractal dimensions between one and two, suggesting that they do not fully occupy two-dimensional surface land space. Batty stops short, however, from realising that roads and streets have a pivotal role in defining this fractal dimension of the city.

To be clear, a road or street may be exactly the same physical width in *Real Space*. The statement that such road is closer to being a one-dimensional object and such street is closer to two-dimensions is intended to be interpreted in terms of how the road or street in question influences the urban map around it; the road will reduce the degree to which the surrounding built city occupies surface land space, whereas the street will naturally act to intensify the use of land. This will be explained in more detail through this chapter.

Urban Vectors

The line of thinking introduced in Chapter 2 may be esoterically interesting, but can it be operationalized to inform urban design? The following represents a tentative proposition. It leads to some insights into different urban forms. If it proves useful, in time it may be possible to quantify this approach through research in such a way that it can genuinely and robustly inform design to give rise to urban forms that can predictably function as planned – for example, intentionally cohering communities instead of inadvertently dividing them. If successful, it may be possible to build algorithms for streets to enable computer modelling, which could counter-pose or be embedded into those models for roads.

The suggestion here is to use vectors to indicate what may best be described as the *focus of attention* of those around a road or street, either on or off the highway. The vectors provide a way to appreciate the difference between *Real Space* and *Experienced Space* (as defined in the last chapter). Figure 3.1 provides the simplest depictions for a typical road or street. These simple sketches can then be extended to deal with more complex urban structures.

The black solid arrows indicate the focus of attention of those on the highway. People on a road are focussed on travelling along it, along the line of the road, in one direction or another towards a destination; further, their focus is primarily on that which is happening on the road, to maximise the speed of their passage. Therefore the vectors from either side of the road are inward looking; the bigger and faster the road, the stronger these inward looking vectors become, pulling the *Experienced Space* inwards, causing the road to more and more assume the characteristics of a one-dimensional line through space, which in turn influences the fractal dimension of the surrounding townscape (as will be discussed).

The grey dotted arrows indicate the direct impact on the landscape around the road; they show how the presence of the road acts to pull apart that which lies on either aside. They show the direction of tension in converting one's comprehension of the landscape from *Real Space* into *Experienced Space*, as a result of the presence of the road. By reducing travel time between the ends, the two destinations seem to be closer together, and vice versa for any proximate objects on either side, which instead experience greater separation.

Looking at the whole set of urban vectors for the road, it can be seen that the side of the road is, in essence, an unstable location where all movement and human attention will naturally move away from that boundary line. This helps to intensify the sense that road edges represent impermeable lines. The hard-shoulder of a motorway, for instance, is a no-man's land, where no one ever wants to be (and that's not just because it is a dangerous location); it is simply a space

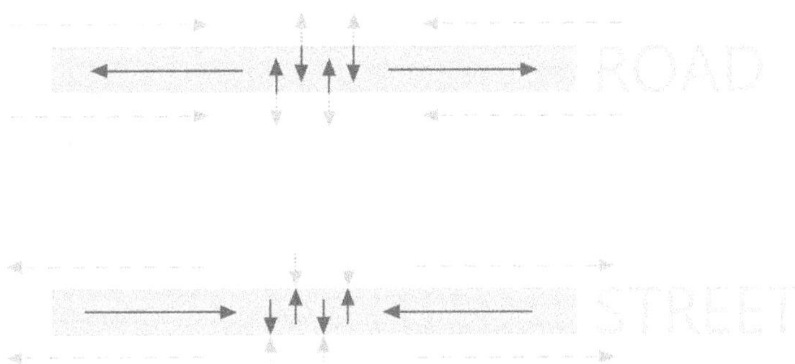

Figure 3.1 Road and street vectors

or place to be avoided, because of how the surrounding space is being used by everyone else.[2] This *pushing away* from the road edge influences the surrounding townscape to move away from, to retract from, to be pulled back from, the sides of the road. The direct consequence is that the immediate townscape to either side is either actively designed or inevitably becomes lower density (or, at the very least, less intensely used) than it might otherwise be. The path of a road through town thereby influences that part of the town to adopt a lower fractal dimension – closer to one-dimensional space. Surface land space around a road is, therefore, naturally less active than it might otherwise be.

The above behaviour of the townscape can be seen to have been and to be taking place throughout many big cities, as they grow and embrace the automobile. There are, for instance, many examples in London where once thriving high streets, intense with human activity and interaction, have become deserted, boarded up, desolate strips of once prosperous shops, as a consequence of the very large increases in traffic rumbling along them, intent on getting to and from distant destinations without a thought for the local (what was at one time) busy high street and urban centre.

At the limit of the busy motorway (or for that matter railway line), the actual surface area of the road is entirely divorced from the urban and rural landscape around it; from the perspective of the local, it is as if that surface area simply does not exist; the edges of the road become completely impermeable lines and the entirety of the road itself is void space creating an impassable boundary cutting through the landscape. Furthermore, such motorways tend to be as straight as possible, to connect major urban centres; consequently the analogy (or perhaps reality) of the fractal dimension of the road collapsing towards one-dimension is very real in terms of how humanity treats, experiences and uses that physical space: our social construction of that space suggests that it has only one dimension, even though we objectively see it as something with a real physical width.

For the pure street, all vectors point in opposite directions. This time the black arrows across the street point outwards, indicating the attention of the shopper on the sides of the thoroughfare, and the end arrows point inwards. The effect of this is to accentuate the impression that the street is a two-dimensional space; when the retail architect has been successful in his task, the shopper will feel that the mall is squarer (wider and shorter) than it actually, physically is; the retail developer further promotes this sensation by book-ending the street with the visible anchor store shop fronts, thereby creating a clearly bounded area; this makes us feel like we are in a two-dimensional space, as opposed to being on a line. For the street, the grey (dotted) arrows show how it pulls together, stitching like a seam, the townscape on either side, yet simultaneously appearing to push apart either end.

---

2 This infers some interesting insight into how our individual perception of space is influenced by how others use space around us. The idea of urban vectors touch upon this phenomenon, which is ripe for further research and investigation. This is not explored in any more depth here.

The combined effect of the end arrows in the street context is particularly interesting. The black (solid) arrows within the street itself suggest a compression of the sense of space to accentuate busy-ness and congestion within that space; they indicate a draw of people into that environment. In turn, the grey (dotted) arrows show the influence on society to treat that space off the street along its edges to be larger (longer) than it really is; in other words, we are collectively incentivised to think that we can fit more into that space than perhaps we can. In economic terms this directly translates into treating such land along the edge of the street as higher value and to seek to build to higher densities within that zone. We collectively try to cram such space, along the edge of the street, full of stuff. The outcome is that the street creates a feedback dynamic that promotes its own continued existence as a bustling hive of activity. And, of course, higher density represents greater intensity of use of space, equating to a city that is closer to being two-dimensional – maximising the use of surface land.

These pulls and pushes are invisible and subtle; but they are quite real; they do influence our individual and collective behaviour, how we value land and how the townscape consequentially evolves over time. They emerge from the urban configuration and manner in which the physical environment is being used, which influences our perception of space and creates a tension between *Real Space* and *Experienced Space*. In the presence of a good, effectively functioning street, two houses in the town (see Figure 3.2), located in the hinterland just off the street, will *feel* closer together for their inhabitants than they really are. That is to say that the occupants will act as if they are closer together by dint of the level of interaction occurring between them (at least, at a statistical level as deduced by Appleyard and Hart in their various pieces of research). A good street promotes interaction between

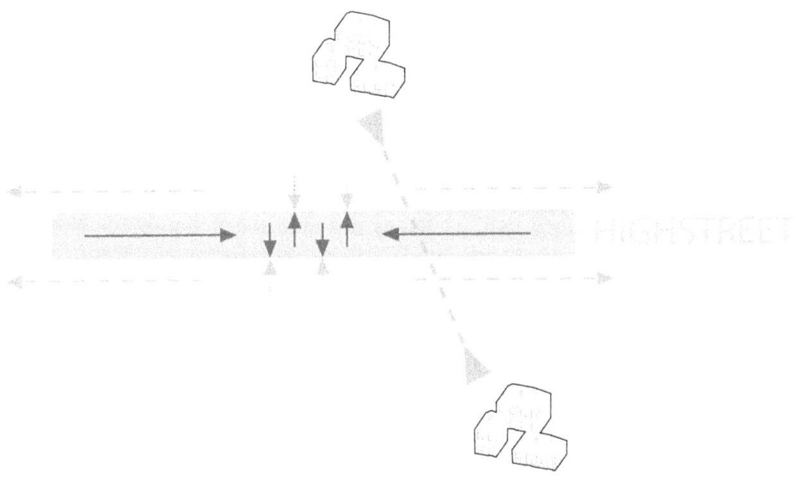

Figure 3.2 Street within a town – connecting across its sides

the hinterlands on either side; a road has the opposite effect. In the case of streets, it is a very subtle effect, but, as will be seen, essential for creating community cohesion.

In terms of the edge of a well-functioning street, shoppers are induced to linger in this zone, reinforcing the sense that street sides are highly permeable. In contrast to the hard-shoulder of the motorway, the edge of the street represents an attractor space, drawing humans and human activity to it.

Quite clearly, the modern out-of-town shopping centre does not confer any of the benefits of a street onto the wider landscape. The modern retail developers have harnessed these natural, social behaviours and responses of humans and replicated them inside a black box (often quite literally) for a corporation's commercial benefit. How that black box itself impacts on the wider urban landscape will be explored in the next chapter.

Having set out the basic principles, we can now begin to see the effect on more complicated urban forms. The two most important and common correspond to the crossing of roads and crossing of streets. These are considered respectively below and in more detail in Appendix A.

Crossing of Roads

When applying the vector approach to a cross-roads, the first point to note is that the junction is not in itself a destination: the objective of the traveller is to navigate the cross-roads as fast as possible and travel onwards. Such junction of roads is either something to be navigated along a traveller's path or simply a bend in their route (see Appendix A). Figure 3.3 shows how the consequential amalgam of vectors creates something with which we are all very familiar: the roundabout. As a consequence, like the sides of a road, the very centre of the cross-roads becomes a location which is naturally avoided; it is visible, but inherently inaccessible. The natural engineering response is, therefore, to design the central space so as to be devoid of human interaction and activity – *something to be seen, but not accessed.* Whilst it may technically be public space, no member of the public ever ventures onto it.

The next natural design response is to funnel the road mouths in such a way that they draw travellers away from such an unstable point in the road network, as fast as possible – speed can, as far as possible, be maintained. The traffic engineering community have invented various reasons, such as the need for sight lines, to justify within design guides why road junctions need to be curved in this way; but all these excuses come down to a simple and singular objective: to maximise speed and flow rate of traffic, thereby creating an entirely void space in the middle.

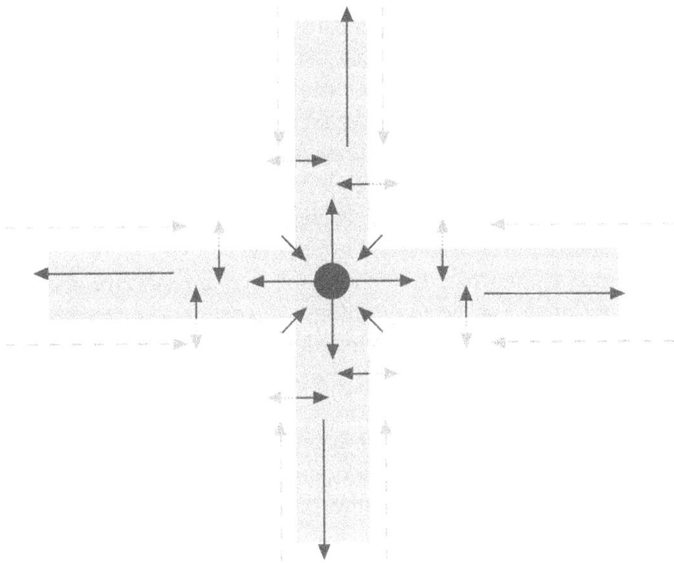

Figure 3.3 Urban vectors for a cross-roads

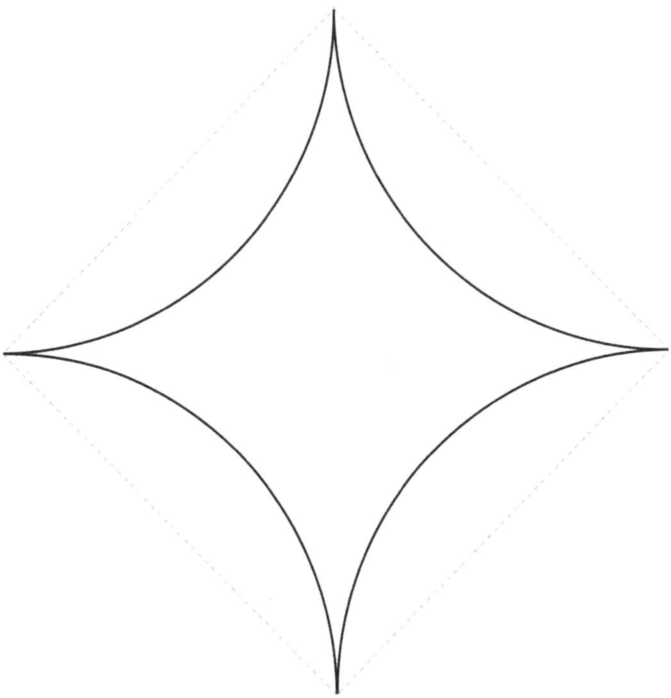

Figure 3.4 Natural shape of a cross-roads

Another way to view the natural construction of cross-roads, with which we are all very familiar, is to appreciate that the inward curved diamond shape is the shape that, for a given circumference, occupies the minimum two-dimensional area (see Figure 3.4). It is as if you have taken a square, with the corners aligned down the roads, being pulled by the tensions towards the four destinations and then sucked all the air out, thereby pulling the sides in. This fits with the idea that the pure road is a single dimensional object: where two roads cross they *try* to occupy as little two-dimensional space as possible.

The cross-road is, in urban design terms, a vacuum space or perhaps it could be called an *anti-place*. In the terminology of Real and Accessible Spaces, the centre of the cross-road might be a location in Real Space, but it is far from Accessible. It is empty space within the urban landscape, a node where no activity happens, where everything goes around and passes by. It is a locus on the map that is devoid of any human interaction. In addition to this, the centre of the cross-road or anti-place acts on the one hand to maximise the connectivity between four destinations and on the other hand divides the local urban landscape into four separate territories.

The terminology used in the above paragraph is knowingly emotive with purpose: to make it loud and clear what the purpose and consequential impact is from crossing roads. This is not to say that cross-roads are bad; they are absolutely essential features within the town and wider landscape. But in serving one purpose, they have very clear consequential impacts on the local townscape, which need to be taken into account in any urban planning.

Crossing of Streets

We are all very familiar with the outcome for road crossings because of the degree to which urban design has become dominated in recent decades by traffic engineering techniques, all focused on maximising traffic speed and flow rates. Using the vector approach for two streets, a very different solution emerges.

Firstly, and critically, the very action of crossing two streets can be seen, in itself, to generate a destination within the urban landscape, where none might have existed before. Focus of attention of those on the accessing streets naturally converges towards the centre of the cruciform. The outcome is that the central point of crossing acts to draw people towards it from all directions. It becomes a point of maximum dwell-time, where people naturally linger. Socially and economically this creates (in a sense) a high pressure point in the urban landscape, a node into which human activity is drawn, focussed and concentrated. It is the complete opposite of two roads crossing. It is a *place*.

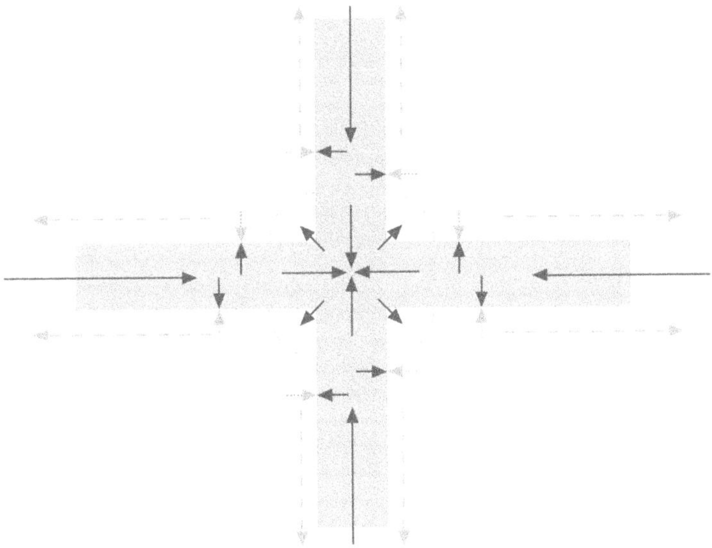

Figure 3.5 Urban vectors for a cross-streets

The natural shape of the cross-street is one which maximises area of public realm (surface area of the space) for any given length of sides (circumference). The outcome, intuited by many historic designers, is that of the circus (good examples in London include Oxford Circus and Piccadilly Circus[3]). Such circular space is designed to hold people within it, with minimum sized exits, which do not automatically funnel people away from the space. Rather the reverse, the streets will direct people into the space creating what is effectively a high-pressure point in the urban landscape, one which naturally becomes congested. Over history many other solutions have emerged, where streets serve circular, square, rectangular, diamond-shaped or other spaces. The circus, however, is the most natural and most space effective solution: for cross-streets think high pressure, blowing up a balloon, whereas a road crossing is a point of low pressure, air being sucked out.

3 Whilst Piccadilly Circus may not be a particular good circular example, it operates as described, being the junction of several good streets.

Figure 3.6 Natural shape of a cross-streets

To repeat, for good measure, a crossing of streets is a node within the urban landscape, which acts as a high pressure point, to which humans and human activity are naturally attracted as a consequence of the design and structure of the surrounding urban landscape. Furthermore, it is a construct within the urban landscape, which stitches four areas of a town together. This manifestly equates to that nebulous concept, which all urban designers of the modern moment aspire to create within their masterplans: it is a genuine *place*.

Where this approach begins to become even more useful is in terms of suggesting where to locate activities within the public realm: for example, it helps to explain other rules of thumb from the retail world. When a retail developer produces a cruciform Mall, then the highest rent locations are situated on the cross itself – location of maximum footfall (of the right type), where there is naturally also a maximum dwell-time. In addition to this, the well-designed Mall is set out in such a way that food and coffee shops can be accessed easily and directly out of the centre of the cross or the middle of the circus (for example through the use of escalators and elevators taking people up out of such space). The clever retail architect constructs his scheme so as to focus people into this central attractor space, creating a higher pressure of humans and human activity, and then uses that high pressure to direct people to other retail outlets, ones which require higher dwell-times such as cafes and restaurants.

These influences of the urban form on the movement of people are all very subtle, but absolutely critical for the successful operation and maximisation of commercial gain for modern retail centres.

Roads and Streets

When this approach to urban design is applied to the wider townscape, then it becomes apparent that the highways around the town should be seen as a flowing system, which is punctuated with pauses. *Places*, those nebulous yet coveted aspirations of all those placemaking professionals, manifest as those locations in the landscape that correspond to such pauses. They are nodes to which humans are naturally drawn, where speed of travel drops away and becomes lateral to the direction of the highway and there is a higher dwell-time. The consequence is a high intensity of human activity, drawing people in, to linger in one place. *Places* happen naturally within the urban landscape as can be understood and explained by use of urban vectors.

This depiction of the urban landscape is a far cry from the modern transport models, which seek to optimise the road network, minimise dwell-time of traffic (of all types) in any one location and maintain maximum flow rates, everywhere. It is no wonder that the placemakers and the traffic engineers are perpetually at loggerheads and fail to understand each other. The trouble with the conventional traffic engineering approach is that longer dwell-times are only considered acceptable at precisely defined *private destination*[4] points (at home, in the office, at the business park, in the Mall or at the cinema complex). There is no scope in these transport models for longer dwell-time in the public realm itself, actually on the streets and at circuses, where *public destinations* (attractor points) arise incidentally out of the junctions of streets, rather than being predefined commercial, privatised entities (as listed above).

A good way to visualise the real challenge facing traffic engineers is to consider a railway line. The railway can be conceived in vector terms as long stretches of road, punctuated by short streets (the stations and platforms) (see Figure 3.7). Between the stations, there is obviously longitudinal motion along the line of the track; at stations, being natural pauses, passengers embark and disembark, creating lateral movement. Minimising travel time between larger destinations at either end of a railway line can most easily be achieved by shooting through the intervening stations without stopping. Instead the system has to be optimised, allowing it to serve intermediate destinations, which requires account to be taken for a proportion of the trains slowing down and pausing to drop off and pick up passengers.

Optimisation can be relatively easy in the case of a single dimensional system such as a railway line; the algorithms are not overly complex. But it becomes horrendously complicated when a two-dimensional network needs to achieve the same effect (fast bits and slow parts). For simplicity (together with a complete failure to appreciate the nature and importance of streets), traffic engineers have resorted to seeking to maximise flow of traffic through the whole highway network – pauses have been eradicated. Gradually they have worked to tweak

4 The concepts of *private destinations* and *public destinations* are considered further in Chapter 4.

the road system, iterating towards faster and faster carriageways everywhere, with potentially disastrous consequences to the healthy and optimal functioning of our towns and cities at the local level, especially our suburban and market town centres.

Taking the use of the vectors a step further, contour lines can be added (as shown in Figure 3.7), which provide a sense of how the combination of roads and streets influence the pattern of human activity across the geographic landscape. Closely packed contour lines indicate points of high intensity of human activity, with streets drawing activity to their sides and roads pushing activity away. If we lose our streets and everything converts to roads, then there is an automatic tendency both towards lower density urban forms (and all the associated implications to be explored in later chapters) and to a loss of that orthogonal direction of activity, which is promoted by streets. Streets enable lateral motion across carriageways, feeding through permeable street sides to promote the finer grain of urban form, those smaller residential streets, lanes and alleyways which connect the larger road network into the local parts of the city.

Figure 3.7 also describes well the historic roads/streets that feed outwards from the centre of a city such as London, as mentioned earlier: Camden High Street, Islington High Street, Clapham High Street and numerous others are all pauses along what have become arterial roads. These carriageways are all roads into and out of town, which are regularly punctuated by streets, which represent the locations of high lateral (as well as longitudinal) activity. These are the streets that have survived. There are numerous others, which, with the growth in traffic into and out of London, have long ago died and become boarded up no man's lands as the streets have incrementally and inadvertently been converted to fast-flowing roads.

The urban vector approach, described in this Chapter, suggests how roads and streets affect the urban environment around them. As you stand in the middle of a road (beware the traffic), it very clearly occupies two-dimensional space; it has a

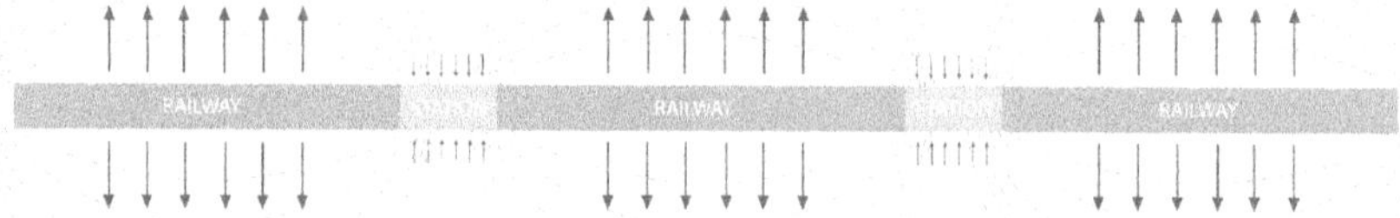

Figure 3.7 Urban vectors for a railway line

clearly measurable and fairly constant width. But the faster and more furious the traffic hurtling along that road, the more the urban vectors suggest that it should be treated as a single-dimensional object. It is not that the clearly measurable width of the road collapses to nothing; rather the manner in which we collectively treat that space means that the road causes the overall fractal dimension of the adjacent urban grain to reduce to occupy less and less of the two-dimensional landscape. In contrast, the presence of a street induces the city to make better use of the landscape and to become closer to a two-dimensional object, occupying and making good use of the whole geographic surface available to it.

Whilst streets help cities become much more efficient and effective in the use of space and enable much higher densities of human habitation and intensity of activity to exist, we must not forget that roads are equally important for the macro-economic functioning of the city. Roads and streets need each other. But it is only streets that have the capacity to create *places*, *public destinations*, on the map of the town.

Chapter 4

Competing Spaces and Urban Dynamics

Since the work of Adam Smith, some several hundred years ago, the idea that there exists economic competition between different geographic regions has become quite familiar. At a regional level it is now well rehearsed within disciplines, such as Regional Economics, how larger cities compete against each other on the national and international stages; it is a constant pre-occupation of the senior politicians of regional cities. Competition across the landscape is not, however, constrained to the macro-scales; it operates at all scales. To understand the influences behind urban planning and urban design, it is essential to appreciate how such economic competition and in turn cooperation take place at more local levels. This provides a much better appreciation as to the inherent differences between *public destinations* and *private destinations*, the concepts introduced at the end of the last chapter.

As introduced in Chapter 1, in the UK and the US we are now being forced, rather uncomfortably, to face the rapid decline of many of our historic town centres, both market towns and smaller suburban town centres. These are often places, which certainly in the UK and particularly in the case of the market towns, have thrived as centres of public activity for many hundreds of years. And yet, in the space of less than a couple of decades, we are seeing them die in their droves. There has been plenty of recent research into this matter, for example that relatively recently carried out for the UK government by Mary Portas (2011), leading to suggestions on how to revive these once thriving centres; but without understanding and working with the underlying urban dynamics, any strategies going forwards will unlikely deliver any sustained revivals.

As will be explained through this chapter, the cause of the demise of our precious town centres (at least in the UK context) has been the result of a combination of an increasingly car-borne society and a period during the 1980s and 1990s of overly lax planning laws. And now there is the internet to contend with too. In retrospect the processes impacting on our dear market towns and suburban centres was highly predictable; accordingly planning policy has been rapidly tightened up; whether the damaged market towns will ever recover proves to be seen and will probably depend very much on the correct application of appropriate planning policy at a local and regional level to slowly counteract the decline; but, as will become appreciated in this and later chapters, in this regard the town planning system is fighting a battle against the natural economic forces. On a more positive note, the experience of the last 50 years and the emergence of the motorcar does, however, provide the opportunity to gain a better insight into urban dynamics and how these operate at different scales and densities of urban form; looking forwards, in timespans measured by decades and beyond, this should help us to begin to plan for a more sustainable future urban form.

How to Kill a Market Town

Reverting to the natural town described in Chapter 1, it was asserted that the structure of such town represents the most energy efficient configuration for the different land uses, where such uses of land generate destinations (usually, but not exclusively, built structures) with identifiably different trip and dwell-time characteristics. Each house and each shop represents a micro-private destination on the urban cloth. Shops become focussed at the location of maximum accessibility in the centre of town, gathering together to create a greater destination, which everyone in a town can reach by means of short, daily visits. Arising from collective cooperation (as explained further below), the town centre represents a *public destination.* For one of our ancestors living in a town, the only place to go to obtain daily food was the town centre; choice lay in which shops or market stalls to visit during a shopping trip; there was, however, no choice in overall geographic destination – town centre, and town centre alone.

The advent of the motorcar has changed all that. Such existing land use patterns, which had been typical for all towns and cities for thousands of years no matter where in the world, were thrown into disarray, for one simple reason: the motorcar suddenly gave people an option to choose where to go when they left the front door of their house. Once a critical mass of people could drive, then all of a sudden it became possible to locate a new group of shops somewhere other than the historic town or suburban centre. From the retail developer's perspective, and the occasional lucky farmer, this was a goldmine; existing land values on the edge of or out of town were proportionately very low, determined on the basis of agricultural productivity; locate a new shopping mall on the erstwhile farmland and the mark-up in land value can be immense. By persuading enough people to drive out of town to the new location, the rents that could be commanded from tenants would be as good as those in the town centre. In paying only a nominal charge for the land itself (unlike seeking to purchase existing high value land in a town centre), for the opportunistic investor these early shopping centres gave rise to massive super-profits.

A whole consultancy industry has grown up around retail development, which seeks to analyse catchments, the geographic area of potential shoppers who might be persuaded to visit a new shopping centre or out-of-town retail park. The factors that play the most important part in any retail catchment analysis are the socio-economic status of the relevant communities (leading to estimates of their likely spending habits), distances (where the most important factor is travel time and not actual distance) and predicted future economic growth. For reasons, which will become clear, numerous retail centres have been justified through planning applications on the basis of spurious overly optimistic predictions of economic growth, thereby suggesting that there will be plenty for all and that the new shopping centre will not unduly detract from those existing town centres in the vicinity.[1]

Once the new shopping centre exists, then each and every time that any potential shopper in the local population leaves their front door and climbs into their car, they can choose: do I drive into town, and have difficulty parking my car and pay for the courtesy, or do I drive to the gleaming new out-of-town, indoor, air-conditioned shopping centre or supermarket with plenty of free car parking? Once planning permission had been awarded for the new out-of-town store or group of stores, the fate of the existing town centre was sealed. The death was not immediate; it takes time for people to completely change their habits, and there is another factor. The demise of these market towns has typically taken around or a little over 10 years to reach a new equilibrium. The timescale for this decay links neatly with the typical length of retail leases – frequently 10 years; once the out-of-town centre had been opened, no major retailer would sensibly renew his lease in the existing town centre. As shops gradually vacate and relocate from the old, when leases expire, then shopping habits of the population become locked-in to the new.

In the UK this process has been, to a degree, nipped in the bud (albeit not before serious damage has been done in many places). But in the US, with its frequently lax planning rules, the new shopping centres have soon succumbed to exactly the same fate. A few years later, in any particular location, another developer talks to another farmer and sooner or later another sparkly, new Mall appears. Now the local shopper has a three-way choice: the now tired and decrepit town centre, the not-quite-so shiny retail mall (visited so often with some retail tenants still selling yester-years fashions) or the very new gleaming shopping Mall with all the new trends. The shoppers, unsurprisingly, flock (genuinely flock) to the new. The not-quite-so new Mall of, say, 10 years of age, is very quickly redundant. And it dies. It is not perhaps surprising, therefore, that the knowledgeable retail developer will only invest the bare minimum in build costs when constructing new retail development and why Malls often look very cheaply fabricated; in places with a liberal town planning system, the *profitable lifespan*[2] of any new retail development is very, very brief.

---

1 This assertion is made on the basis of anecdotal observation working within the property industry and in particular through experience promoting major new retail developments.

2 A reasonable revenue can still be generated (at least for a while) by an out-of-town shopping Mall after 10 years, but under competition from other newer Malls the return on the investment will continually decrease, approaching background levels (i.e. bank interest rates or below). This clearly does not apply to those few lucky investors in out-of-town Malls in the UK (such as Bluewater (Kent) and Cribbs Causeway (Bristol)), where subsequent changes in the planning system have frozen the geographic landscape and prevented new competition from being built.

It is hopefully self-evident from this recount that when people can choose geographically where to shop, then those destinations are operating in direct and absolute competition with each other. Only in exceptional circumstances will anyone drive from one shopping centre to another. Most people (essentially all people, everyday) will choose to visit one or another on any one day. In economic terms, each destination is operating exclusively of any other; a shopper will spend money at one or another, but never both. All *spend* diverted to a new centre is lost to an old centre. Unless there is significant population and economic growth (as necessarily predicted in all those retail impact assessments accompanying new planning applications), then any new centre will inevitably cause a loss of economic activity in any older centres. In turn, a reduction in economic activity results in a fall in land value. The old decays and dies, forever to be out-competed by something newer and shinier. The stark evidence of this process can be seen all over the US; it has even been documented in a dedicated website (Dead Mall Stories, 2014).

But not everywhere has suffered the sorry fate of the market towns or the smaller suburban centres or those once shiny new Malls. Many big city centres have bucked the trend and not been damaged in the same way. To appreciate why they have survived requires a little exploration into agglomeration economics and how this manifests spatially.

Very Localised Cooperation and Competition

Reverting back to the natural town again, it has already been asserted that from the wider population's perspective having the shops in the middle of the town represents the most energy efficient structure for the town. There is, however, an alternative way to appreciate why the shops all aggregate at the centre. The question needs to be asked: is it the best solution for each individual retailer? If a shop were to locate itself away from the centre, perhaps halfway out to the edge of one side of the town, surely it would provide easier access to those homes immediately around it and thrive accordingly? Why, then, aren't the shops servicing a town evenly dispersed throughout the town, each seeking to be as far apart as possible from all others?

The classic example, which is used to demonstrate that the above is not so, is through a thought experiment about two ice cream sellers. Imagine a beach, perhaps a mile long, with a road and car park running along the back of the beach. On this imaginary beach the long car park behind means that people are evenly distributed along the whole beach, rather than bunched around an entrance. Now, imagine two ice cream vans arriving at the beach to sell ice creams to the throng of people sunbathing themselves and swimming in the sea; perhaps one ice cream seller arrives first and parks at one end, such that when the other arrives he sees the competition and, being seen as competition, he automatically parks as far as possible away from the other, at the other end.

As the day progresses, perhaps it cools and some people leave the beach, so business slows a little. Competition thereby increases. The newcomer thinks, 'if I drive a little along the beach then I will be a shorter walking distance away from all those other people at the other end, who might then decide to come to me instead'. So he drives a hundred yards along and continues trading. In the meantime the other sees his competitor approaching and realises that he must do the same. Each time they move, they are making a rational economic choice to maximise their potential retail catchment and thereby potential trades and profit. They consequently each iterate towards the middle of the beach in their attempt to improve their personal profitability and out-compete the other. Eventually, the logic of the argument goes, they both end up parked right in the middle of the beach each most accessible to the most number of people. Despite being competitors, based on rational decisions they have each ended up located right next to each other (a little counter-intuitively, perhaps).[3]

As the two ice cream vans approach the centre point, they need to find some other way to differentiate their product to entice different purchasers of ice creams to buy from them instead of the other. If a third ice cream van were to arrive, then following the same logic, he too would end up at the centre of the beach, directly alongside his competitors, differentiating his product in yet another way – perhaps one selling chocolate, one strawberry and one vanilla ice creams. The apparent outcome from this process is that the competition for customers has driven the different vendors to cooperate in the physical world by co-locating at the centre of the beach, thereby creating an ice cream *destination* at that place. And so it goes with all the shops on the high street. For many (if not most) goods, it is better to be physically co-located and then seek to differentiate products and fashions and brands, than to be isolated at a distance and never have the opportunity to persuade shoppers that there could be a better alternative.

In answering the question at the start of this section – 'what is best for the retailer?' – at sufficient housing density there is indeed scope for a generalised corner shop or two at locations other than the centre of town; but for most specialised retailers, or any shops where customers might want to compare purchasing opportunities, then being in the centre with everyone else is by far the best location to be.

Economic Space

This brings us to a new kind of *space*.

3 Note that this course of events is driven by competition, where the more intense the competition the more it will drive the competitors together. On a gloriously hot weekend day, when the beach is packed, there may be more than enough business for each van to remain in its original position throughout the day; in this case, there is less competition and the incentive to move to the middle ground is diminished.

In Chapter 2 the concepts of *Real Space*, *Experienced Space,* etc. were introduced. There is another type of space, which also needs to be accounted for: this is *Economic Space* (it could equally be called *Social Space* or *Cultural Space*). This *Space* has no formal connection to the physical landscape; rather it is a social construction of human society. If it makes it easier to understand, then think of the Internet as a direct physical manifestation of such *Economic Space*. It is a virtual space.

Clearly all retailers are competing directly for customers and thereby for revenue; any individual choosing to buy a particular good from one retailer is generally choosing not to buy a very similar good from another retailer or to have less money to spend on something else. All retailers are, therefore, always competing in *Economic Space*. But, just as in the real world, retailers can be closer or further apart from each other in *Economic Space*: two shops selling exactly the same goods would be very close to each other in this virtual world, whereas two shops selling distinctly different goods would be a distance apart. The further apart two retailers are in *Economic Space*, the less the perceived competition between them because people may likely need both types of shops in daily life. A food shop and a hardware store may still be competing for a person's monthly budget, in that buying new tools may reduce expenditure on food, but not by much (a family is likely to have – even if only notionally – a budget for food and a separate budget for hardware). Two different bakers, on the other hand, are very directly competing.

Competition in *Economic Space* then becomes expressed in terms of locational choices in *Real Space*; retailers who are closer to each other in *Economic Space* will prefer in first instance to locate at a distance from each other, so that a visitor to one would unlikely visit the other. This is fine when the future is optimistic and where there is plenty of *spend* within a catchment. But, as seen with the ice cream vans, where competition becomes increasingly intense (more retailers or fewer customers or customers wishing to spend less), then this drives retailers together; when competition is intense, it is better to have the opportunity to persuade potential customers to drop by your shop and perchance buy a product than never have that opportunity in the first place. When the retailers are driven to co-locate geographically, they are then forced to identify ways to differentiate themselves and thereby seek to distance themselves within *Economic Space*. From the retailers point of view, it is not a happy circumstance if they are proximate in both *Real* and *Economic Spaces* – one or the other, but not both.

This can be seen to be played out in the location decisions of different retailers. For example, out-of-town supermarkets, Malls or the department stores within Malls – all being generalist food and clothes retailers – will always endeavour to distance themselves from others of the same, when they can and when there is plenty of *spend* within the catchment. In lax planning system environments, they will all tend to pick different geographic locations, so that potential shoppers have to choose on leaving their home which one to visit. When either the travel patterns of potential shoppers (on foot or reliant on public transport) or the planning system forces shops to concentrate in town centres, then these generalist food and department stores will intuitively seek to distance themselves (say, locating at each end of the mall or high street[4]). The reason for seeking such geographic separation is that in being generalist they find it difficult to differentiate themselves in product terms: all the food shops still sell Heinz Baked Beans.

Where, in contrast, there is natural distance in *Economic Space* between two retailers, they will happily co-locate (for example men's verses women's clothing), thereby creating a bigger destination to attract customers without themselves overly worrying about having to compete for the spend of those visiting customers. Rather, they are likely to gain spontaneous additional customers – say, a woman going clothes shopping for herself may take the opportunity to drop by a men's store to buy a birthday present for her partner without it costing her much extra effort or time to do so.

When, however, competition becomes very intense, either for specialist goods, for which there may be relatively few customers, or when customers generally are spending less, then the dynamic evolves. It becomes increasingly important for retailers to aggregate together, to have the opportunity to capture any possible additional expenditure from spontaneous customers or by proving that their products are better quality or cheaper than the competition. This drives the shops, even if they are ostensibly similar, together.

In co-locating, two or more retailers are expressly coordinating their choice of location to allow people the opportunity to visit both shops before making a decision on what to buy. In the retail world this is known as *comparison shopping*, where shoppers are able to compare products before purchase. This contrasts to what is known as *convenience shopping*,[5] where shoppers do not bother to seek to compare products before purchase (for example, when you go to a supermarket, you have (probably) forgone the opportunity to see what price and quality the same or similar goods might be at another supermarket).

The same logic applies to the overall structure of a town as much as to the internal structure of the supermarket or department store. Apples are co-located with other apples to enable shoppers better to choose which apple to buy. In the department store all electronic goods are located together, as are kitchen utensils,

4 One apparent exception to this rule would be the siting of the departments stores along Oxford Street. This can be explained by reference to the time and nature of these shops when they were first established. The department stores were a new phenomenon and they were competing for a much reduced population of customers. The intensity of that early competition drove them to choose to co-locate, where now they would never dream of doing so, but are obviously stuck to those historic decisions.

5 Whilst retail planning consultants (in the UK, for instance) will reel off a list of items considered to be comparison goods (say, clothes and jewellery) compared to convenience items (say, food and basic toiletries), in reality the difference depends upon the wealth of the shopper(s) in question. At a certain level of affluence it is no longer useful time spent to compare the prices of everyday items in different shops. But someone who is very poor, needing to watch every penny, may still consider it fruitful use of time to go to a food market so that they can be sure of buying the cheapest apples.

and so on. And within the shopping centre those clothing brands focussing on a similar population catchment (girls, boys, men, women, rich, poor) will tend to be collected together into groups.

In the last chapter, consideration was given to the typical structure of retail Malls, where anchor stores are positioned at the apexes and all the other shops line the streets in-between. This fits exactly with the description above. The anchor department stores are generalist shops, which seek to cater for all potential shoppers – a bit of everything is their general motto. It therefore makes sense that they are located as far apart as possible in order to minimise the likelihood (or frequency with which) shoppers will visit both or all department stores within a Mall in any one visit; they are directly competing in *Economic Space*; their proximity in *Economic Space* is expressed as distance in *Real Space*. In contrast all the other shops, which locate along the sides of the malls within the Mall are cheek-by-jowl situated next to each other; in this respect the various clothes and other specialist shops are manifestly cooperating physically in *Real Space*, while still necessarily competing in *Economic Space*.

Whilst the Malls may be artificially conceived environments, the same process arises in towns, which have grown organically over time. This effect of co-location goes further than the creation of the whole high street, as we tend to know it. It dictates, for example, the siting of collections of different types of shops. Walk around London, for example, and it is often quite peculiar how, all of a sudden, one comes across a group of, say, outdoor clothing and equipment stores (selling ski and mountain wear and so on). Given that they are clearly competitors for market share, it seems counter-intuitive to find them all co-located on the same street. Yet the reason for their close proximity is simply that such closeness benefits all of them by drawing in a larger number of people to that area than if they were each sited at a distance from each other. Together they have created an outdoor clothing destination – a *public destination* – on the map; but in order to survive, they then each need to seek to differentiate themselves from all the others – to find some brand or identity or provide some particular service that connects better with potential customers than that of the competitors.

This is agglomeration economics operating at a micro-scale.

The outcome of this in urban design terms is that we see small huddles of seemingly competing, and yet physically cooperating, shops around the town centre. In one area there may be a group of, say, jewellers, all obviously competing economically, yet oddly enough having chosen to co-locate. Along any longer high street, it may be better for coffee shops all to co-locate at one end or another, than to be dispersed evenly along the high street. In that way any one feeling like a coffee will have to walk past and see all of them at once and then choose which one to stop at, than simply to drop by whichever one they see first and not give the other coffee shops a chance to demonstrate that their service, coffee and coffee accessories might be better or more suited to an individual's peculiar tastes. In Malls the restaurants and takeaways are frequently all put together in a food-market area – as noted earlier, frequently at the top of the escalators feeding out of the main *place* within the Mall.

The degree to which physical cooperation can be observed in any particular location will strongly depend on the overall commercial success of that location. Where there are very high footfalls of pedestrians and where a coffee shop, for instance, can expect to run a bountiful trade regardless, then co-location with other coffee shops is less important. But where and when footfall drops, and competition for a limited audience rises, then it becomes increasingly important for similar types of shops to co-locate. It is quite counter-intuitive, but, as noted with regard to the ice cream sellers, the stronger the competition the more that similar shops are driven together physically in *Real Space,* and forced to seek to differentiate and thereby distance one from another in *Economic Space* (see Box 4.1).

Box 4.1: St Christopher's Place

An excellent example in London of how this works is provided by St Christopher's Place. Just along Oxford Street from Bond Street tube station, somewhere between Selfridges and the other highly visible and resplendent department stores, hidden down an alleyway no more than six feet wide – and almost impossible to notice if you are new to the area – is a small square surrounded by a wonderful array of small boutique restaurants. It is a highly successful eating destination in the Marylebone area, which has arisen quite by chance and seems entirely inexplicable in its existence. Its success is based entirely on these principles that, by grouping together and in aggregate creating a *public destination* to which people will be willing to walk from much further afield, these numerous small restaurants have far better likelihood of success than if distributed evenly across the townscape. They may all directly and exclusively compete in *Economic Space* (you would never buy two lunches in one visit), seeking to persuade diners to pay good money for their particular food, and yet they manifestly cooperate in physical space. In order to make such intense competition in such a confined space acceptable and bearable, each restaurant very clearly differentiates itself from all the others by offering very different food. And in totality a huge variety of food types are on offer making the whole destination a very attractive place to visit – even if it involves a bit of a walk.

The beauty of a place such as St Christopher's Place is that very few developers would have the appetite to risk to create such a venture from scratch. It is something that has arisen naturally out of market forces and the (at least initially inadvertent) cooperation between the restaurant proprietors. And through co-locating, these restauranteurs have caused the local land value around St Christopher's Place to rise significantly above what it might have otherwise been. Equally importantly these higher land values are sustained indefinitely over time, held in place by the aggregation of restaurants and cafes. If any one unit fails, then it is quickly replaced by another food entrepreneur; and the *place* as a whole carries on regardless.

To summarise these observations, in stark contrast to the out-of-town shopping centres competing and causing loss in trade and land value from each other, in other contexts it can be seen that competing economic operators choose to co-locate. And such physical cooperation both builds land value and creates stability. These competing economic agents rely on each other's close proximity for their own trade. The huddle of shops that they create becomes a stable point on the map that can last and last and last. By cooperating and choosing to co-locate, a grouping of restaurants, say, creates a *public destination*. This reasoning suggests that the defining feature of *public destinations* is that they are derived from physical cooperation on the urban landscape, rather than competition. Where competition remains, it has become differentiated within our cultural environment (*Economic Space*).

These microscopic influences are played out on the urban landscape to create the underlying urban dynamics that drive the growth and form of our towns and cities. Furthermore, they represent positive feedback processes, such that they build on each other: the more outdoor clothing stores there exist in a particular location, the more the draw for another outdoor clothing retailer to seek to co-locate a store with them. It is this (counter-intuitive) conversion of competition in *Economic Space* into cooperation in *Real Space* that, playing out over time, has created our greatest cities. Success builds on success, and that, in turn, converts into desire, which translates into land value, which, because it is generated by many separate players, is consistent and stable over time. And the larger and more diverse the city centre grows to be, the more stable and durable it becomes. Our many, now sorry, smaller market town centres were not large enough to be able to survive the impact of a new out-of-town supermarket; in each case they have been out-competed by a new destination on the landscape. But the large centres have survived because they are big enough for the supermarkets to want to be located within them instead of in competition with them (see Box 4.2).

Box 4.2: Croydon Town Centre

It is useful to explore in a more specific sense the processes at play in determining how land value is fortified or eroded. The scenario of competing shopping centres has already been considered. Yet it can also be seen that within any shopping centre individual shops cooperate or collaborate in physical space, to draw shoppers in and generate foot fall and in turn trade. This is ultimately what generates land value: trade generates revenue for the retailers, which pays for the rent, where capitalisation of the landlord's rental stream determines the property value and thence underlying land value. In the retail world a rent is now often directly linked to the revenue generated in a shop.

The case study of Croydon (South London) provides further insight.

Croydon town centre contains an unfeasible amount of shopping floor area. Currently there are well in excess of 300 shops occupying 211,000 sqm of retail space and 47,000 sqm leisure space (London Development Agency, 2009). But the hundreds of shops are not just located along a high street. There is indeed a long pedestrian high street, connecting the entire shopping area. But along this street there are already two large, entirely discrete and separately owned shopping centres, competing economically. At time of writing, a third shopping centre is to be constructed, to be located at one end of the high street. The question to be asked is: what is the likely effect of this new shopping centre on the rest of the high street and, in particular, on the other Malls?

If the proposed additional arcade for Croydon town centre were to be located at a short driving distance from the town centre, such that shoppers would have to make a decision where to drive to, then the answer is obvious: people would have to make exclusive decisions – one or the other. For every shopper choosing to go to the new, a shopper (and all his or her spend) would be lost from the existing. The two separated centres would be competing in both economic and physical terms. And in competing in physical terms, each separate centre acts to erode the land values of the other.

But the new proposed Westfield Mall for Croydon would not do that. Any shoppers choosing to come to see this sparkling new galleria of shops would (depending on where they parked their car) provide additional footfall to the existing shops.[6] A percentage of these visitors would drop into the older shops and buy things. If the new Mall were to act to attract even more shoppers to Croydon town centre instead of elsewhere (say into central London or to Kingston), then the additional Mall could even enhance the trade experienced by existing shops. The effect would clearly be more marked for those shops nearer to the new shopping centre, possibly drawing shoppers away from those shops at the other end of the main high street.

The economic consequence of the appearance of the new Mall would therefore be to raise land values, where the nearer the new Mall the greater the rise. In this respect, the land uses are acting cooperatively to build land value, further reinforcing Croydon as a major town centre on the London map.

6 Note for Croydon planners: location of the car parking is critical.

Patterns of Land Value

Reverting, yet again, to the natural (historically pedestrian) town, the patterns of land value can be seen to map directly against levels of accessibility. The more accessible a location is, measured as number of people (or proportion of population) who can easily reach that location from where they normally sleep at night, then the higher the (potential) land value. In higher density locations, where most people rely on pedestrian, bicycle and/or public transport, then this gives rise to a very recognisable land value map, which in turn broadly simulates Von Thunen's concentric ring theory (though, as discussed, for subtly different reasons). Furthermore, as noted earlier, *public destinations* are stable places on the urban map; this creates sufficient certainty to allow investment in public transport. This further locks in the correlated pattern of accessibility and land values. This will be discussed further in Chapter 11.

At lower densities, the correlation between land value and levels of accessibility remain; those locations that are highly accessible *can* have high land value; but, and this is a very important *but*, as indicated in the discussion above, levels of relative accessibility in the landscape are no longer fixed in the same way. Location and accessibility levels become unlinked. The car opens up all manner of new mobility options for the population; the town centre no longer has any (or a sufficient) accessibility advantage over numerous other locations. Other points on the landscape may be further afield in distance terms, but as previously mentioned, distance is not the issue; rather travel time and perceived effort are the critical factors; in a car anywhere in a 10 to 15 minute drive (or perhaps more) is as good as the historic town centre. The consequence is that land values become unlinked from location; land values become fluid; and in each and every location the value of the land becomes fickle. The implication of this for our economies is immense.

The conclusion from this is that when the population is pedestrian bound and where cooperation in *Real Space* is able to take place, then such cooperation leads to stability and durability of land values across the urban landscape. In contrast, at low densities, when a significant proportion of the population gains motorised mobility, then sizeable *private destinations* are able to arise and begin to compete and erode the land values of each other and those of any existing high land value locations or *public destinations*. Any new entrants onto the landscape directly compete both economically and physically with all extant destinations. The land value map consequently becomes fluid and cannot be relied upon to last the test of time.

In the various reports on how to reinvigorate our dying town centres, which have been published in the last few years, such as that written by Mary Portas, much has been made of trying to find new uses for the town centres or investing in them to improve their appearance and attractiveness or simply providing free car parking. The unfortunate reality, however, is that a spate of liberal town planning during the 1980s and 1990s has given rise to too many destinations on the urban map – too many *private destinations* competing against each other and against the existing *public destinations*. The only thing that can genuinely breathe life back into the town centres is economic growth and population growth (in particular more housing within the towns, preferably to a relatively high density and preferably close to or around the existing town centres) and strict planning rules that prevent any new out-of-town destinations being created. In this way footfall could once again increase, revenues improved, land values sustained and eventually a positive cycle of improvement could be re-achieved. Further positive planning needs to take place to prevent the existing out-of-town centres from expanding or, alternatively, turn them into new town centres themselves so that they have corresponding housing to go with the retail (see Chapter 5 for further discussion on the relationship between different land uses). Without these various measures, any public investment into the decaying town centres will be in vain.

Beyond Retail

So far, the discussion has been mostly focussed on shops. But exactly the same principles, in regard to cooperation and competition in the generation or destruction of land value, can be applied to the other major land uses: commercial and residential. The implications of this will first be considered with respect to commercial premises and business parks.

In the context of a car-born society and under a libertarian planning system, it has already been seen that the average profitable lifespan of a shopping Mall is little more than 10 years – the length of the typical retail lease. Applying the same thinking to business parks, it becomes apparent that in low density, car-reliant areas, the profitable life (providing a return on investment above background levels) of a business park will typically be 20–25 years. This equates to the typical commercial lease length. At the end of the lease, any successful business sited at a business park will naturally review its *position* and assess what new offices it can afford and whether there would be a better place to move to. The probability is that there will be … just down the road, just built and being heavily marketed, right now. The existing, accessible only to cars, business park will begin its inevitable demise; after 20 years, no matter how much tarting-up is done, the place will look tired and it will be expensive to bring it up to scratch with all the latest technology; it will only be suitable to businesses that cannot afford the latest and smartest premises; rents will inexorably fall and it will become harder and harder to justify investment in the up-keep of the buildings and infrastructure.

The advice, therefore, to any investor in commercial property is that, in the context of a liberal planning system and good availability of land, if you invest in out-of-town business parks your investment is on an inevitable downward spiral from Day 1. Its peak value is always on the day it opens! The rate of decline will not be very apparent in the first 10–15 years. But after 20 years, it will be quite rapid. In the UK, the factors that counteract this decay in land value are a planning system that exceptionally limits the supply of land (through strong protection of the countryside and green belt), which essentially acts to hold up the land values of those areas in existing use. Investors will need to be wary if UK planning becomes more liberal again, allowing more development. In many parts of the US, for example, the decay in value of business parks is inexorable and inevitable: every newly constructed business park within a wider metropolitan region will act to detract value from all those already constructed and occupied. In low-density urban areas, all new development continually dilutes existing investments, unrelentingly. Only those investors who manage to remain permanently at the leading edge, putting their money into each new, emerging development, can expect to sustain significant commercial returns. Property in low-density urban settings with liberal planning systems is not for holding onto (except during periods of very, very rapid economic growth … but get out fast before the growth stalls).

The manifestation of this process across the US is that the landscape is becoming increasingly cluttered with dying and dead business parks, to keep company with the dying and dead Malls. In a country where there is seemingly limitless space, then once a business park starts its decline there is no *bottom* to level out at. Land value will fall and fall until it reaches zero, or even becomes negative because it would be more costly to re-use this land than to start again on some greenfield (previously unused) site.

In the UK, with much more limited land available, there is a minimum land value to which the business park can fall. Land is in sufficient demand for other uses that it has an innate value. But at low densities, this innate value cannot be realised simply by trying to smarten up the existing buildings and infrastructure. Rather the only way to raise land value again is to knock everything down and re-start again. We see this happening all over London. Old business parks in relatively low density areas declining to a point that they are useless investments (rents are so low that they become liabilities). No matter how much local authorities would like such places to be brought back into use for commercial purposes, the land is marred. The only solution is to flatten the place and re-build and re-brand from scratch – more often than not nowadays re-building at least part as housing.

Housing

Rather disconcertingly, exactly the same processes also affect housing – only the timescales are even longer. For residential areas, the turnaround is somewhere between the generational gap (say 30 years) and the typical length of occupation of family homes (can be as long as 60 years). Again, exactly the same principles apply. In higher density towns, underlying land values of homes directly correlate to access time to a town's centre. In low-density (car-reliant) areas, there is no longer any correlation between accessibility and any point in geographic space, so there is no longer a correlation between land value and location. Land values, therefore, become both inherently fluid and in competition with every other location.

There results a kind of slow motion flocking effect (fundamentally no different to people flocking to visit a sparkly new out-of-town Mall or companies trying to secure space in a nicely landscaped new business park). As a consequence land values in certain neighbourhoods rise artificially above the background (simply because they are deemed nice places to live and people flock to try to live there) and land values in other neighbourhoods correspondingly decline. The outcome is a patchwork quilt effect of randomly high value areas and low value ghettos, with no logic behind the relative accessibility of such locations.

Because much of London's suburban landscape is at a density below the pedestrian accessible threshold, then this is what has happened. This patchwork quilt of high and low value neighbourhoods then becomes locked-in, not because of accessibility, but as a result of the performance of local schools. The consequence is the seemingly random geographic distribution of very good and very bad state schools, which every year create a headache for those parents seeking to get their young children into their first or second school.

In the UK context, this mosaic of high value and low value locations is held in dynamic stasis because of the general lack of land available, which means that all land does have some value; all land within the metropolitan area is still needed by someone. Areas of lower density housing, which have become lower value, are stuck for eternity as low value areas.[7] It is even worse than the business park scenario, because of the piecemeal land ownership. These areas become tired and there is little incentive for owners to invest in their properties. Anyone *making good* (so-to-speak) is best advised to move to the higher value locations than to invest in their present low value place of living. In contrast the high value areas remain stuck as high value, because there is a continual pressure of in-migration from the low value areas. It is a little akin to the notion of high pressure and low pressure nodes created by cross-streets and cross-roads; instead there are high pressure and low pressure regions of town, where the low pressure parts cannot ever be recovered without some form of major state initiative.

In the US (and also in the north of England) an even worse fate beholds the residential landscape. Initially, there has occurred the same patchwork quilt effect of higher value and lower value neighbourhoods, which have no locational logic to them. But, with plenty of land and a more liberal approach to town planning, there is no residual value to the land at the locations of the emerging low value housing. The downward trend therefore continues unabated with no floor. The bottom literally falls out of the market and the low value residential areas decline and decline until there is no value left; worse they can actually become negatively valued, being more costly to redevelop than greenfield land. In some ways, this then creates an opportunity for a savvy developer to acquire all the land (because the location has become mostly deserted) and then to re-brand the area and start all over again. In doing so, however, such developer creates a sparkly new suburb for people to go to, which in turn may start to impact on the value of any other relatively local area that had previously increased in value. Rather than the urban landscape being held in a permanently dynamic stasis of low and high value, the consequence is that the residential suburban landscape in the US is completely

7 Recent research by Douglas Noble (Noble, 2012) and his team of public health experts has shown how durable this dynamic stasis is. They compared current demographics to those mapped out by Charles Booth in his Poverty Maps of the late 1800s. The surprising outcome is that those areas over 100 years ago, which were poverty-stricken and housed the poorest, remain the most challenged locations with greatest degrees of ill-health and other signs of poverty.

fluid. There is no certainty anywhere that any housing will remain at high value over any extended period of time. Just because it is a pretty high value location now, does not mean that in five or ten or twenty years' time it will remain so.

With regard to the economic collapse of 2008, much has been made about the action of the bankers and the supply chain of salesmen seeking to flog mortgages to low-income households. But underlying all of this was a complete miscomprehension regarding the nature of low density housing as an investment. People were persuaded to buy existing and new homes on the premise of rising land and property values. The latter had been happening because of a combination of rises in wealth (generally) and on-going rapid population growth, both of which were putting upward pressure on all land values (economic and population growth exceeded new house building). The underlying urban dynamic discussed above was therefore hidden – for the boom. But if there were any significant increase in the rate of housebuilding or as soon as there was a stalling in household growth (numbers of new households created) or economic growth, then the underlying dynamics took over: housing values in those less favourable neighbourhoods plummeted. The outcome has been the checkerboard cities of randomly sited very high land value areas and zero land value (increasingly deserted) ghettos.

In the context of a libertarian planning system, a car-borne society and plenty of land and building large swathes of housing at low densities, population growth has to consistently and significantly exceed the rate of housebuilding to keep underlying land values static or on an upward trajectory. Otherwise the natural urban dynamic takes over and land values become entirely fluid. The landscape becomes a seascape, where the savvy household must keep moving to avoid becoming rutted in an area of falling land values and consequently losing decades of hard-earned capital. At low-density, there is no inherent stability.

Density Thresholds

To the casual eye, driving through a town, none of this is very apparent. All that is seen is a fixed, constructed world of bricks and mortar, where some areas look good and desirable and others less so. The changes discussed above, arising from the underlying urban dynamics of cities, and in the context of a liberal land planning environment, take place over timescales to which we individually find it difficult to relate. Typically investment decisions in the business world now take place with little more than two-year time horizons; these are much shorter than the time spans of long-term urban dynamics, even retail dynamics, enabling investors to jump ship when the tide turns against them. Homeowners are not so fortunate; mortgages are typically a long-term endeavour. When the underlying economics turns against an area, land values can drop very fast and the physical deterioration can follow quickly behind; it can easily catch out the unwary private investor. And in 2008 it hit many areas at once, catching out much of the world's banking system in the process.

The message to any investor in housing, as with an investor in retail or commercial property, is that, in low density and liberal town planning contexts, the underlying land values will always be very fickle and fluid. A slight downward trend can rapidly convert into a rapidly deflating spiral of decline, from which the only solution is complete renewal (or rebirth) of an area … a complete rebranding into a new, desirable location. This will likely require complete demolition and starting again. Furthermore, in the context of long-term investment, those checkerboard squares of very high land values are very precarious: just as with the town centres of the UK's market towns, in a car-born world there is no economic rationale for them to remain high value into perpetuity. For any individual location, when decline comes, it could be very rapid. So be warned.

The role that the land use planning system plays in influencing and controlling these urban dynamics will be considered in more detail in Chapter 12.

The observations made through this chapter beg the question: is there a density threshold? At high densities, human activity on proximate areas of land appears to cause increases in land value, where such land values become locked-in, stable and durable. In contrast, at low densities human activity very obviously operates competitively in *Real Space* across the geographic map such that different locations can act to degrade land value from other locations. By deduction there must be some density threshold, below which land uses compete and above which land uses cooperate. In Chapter 6 an attempt will be made to derive this threshold. First, in Chapter 5, a similar conclusion will be reached, but from a very different perspective: looking at how different land uses interact across the townscape.

Chapter 5
The Ingredients of the City

The last chapter focused on how the human occupation of land can give rise to competition or cooperation with other similar uses of land and the consequences in terms of land value: for example shops cooperating (in locational terms) with or competing against other shops to reinforce or erode land value. Through this chapter consideration will be given instead to the locational relationship between differing land uses, in particular focusing on how this varies across urban densities.

Over the last couple of decades there have been some substantial shifts in thinking about what constitutes a well-functioning urban area. One of the major changes has been a promotion of the notion of mixed-use development. This has been a counter-response to the perceived failings of zonal planning,[1] which dominated for much of the previous half-century. The new focus has been to mix up different types of land uses, combined with an attempt to promote pedestrian activity – as we shall see, potentially all in vein.

Since around 2003, a blanket requirement was embedded into UK planning policy to require that any new development of any size (in London often taken to be anything from a small block of flats and larger) should incorporate other land uses into the development. In the absence of any good theory in urban design as to how this policy should be applied, the consequences for London's urban landscape is beginning to prove to be as disastrous as the previous zonal approach. Where zoning gave rise to large housing estates or business parks divorced from any other land use, making it impossible to do anything without jumping into a car, the uncompromising enforcement of the mixed use policy has led to new housing developments across London that are permanently blighted by empty, boarded up space on the ground floor.

---

1 Zonal Planning. For those that are not familiar with the terminology, Zonal Planning represents planning by map, where a town planner identifies different areas of the urban landscape that are suitable for a zoned use – say some businesses here, some shopping there and residential over there. In the context of the discussion in the last two chapters, this represents the creation of artificial *private destinations*. Inevitably such zonal planning depends upon the creation of roads (that is ROADS) to connect the different zonal areas together. This approach to town planning was probably the default approach for much of the latter half of the twentieth century, certainly in the UK and US.

Land Use Ratios

If you take a walk around Mayfair and Marylebone in central London and peruse some of the side streets, it is quickly apparent that, apart from the prosperous shopping streets and a few busier places, very few buildings are vertically mixed use – that is more than one land use in any one building (e.g. homes above shops). Even at the density of such a central London location, most buildings are in fact still mono-use – just housing or just commercial. Taking in this observation, the notion that any off-pitch (i.e. away from the high street) locations in outer suburban London should be vertical mixed-use seems somewhat misconceived. But why should this be so? Why not vertically mix land uses anywhere?

To understand better how mixed-use planning policy should actually be applied requires reversion to some statistics about the entirety of our building stock. This knowledge can then be combined with understanding around movement patterns to enable a better appreciation of how typical ratios of different land uses within urban areas are naturally expressed across different scales and densities of urban development.

Across the whole of England very approximately the ratio of built residential floorspace against the built floorspace of all other urban land uses (excluding farming and heavy industry) is 6:1. That is six square metres of homes for every one square metre of the aggregate of shops, offices, leisure and urban warehousing (see calculations in Box 5.1). At a statistical level all this development area can be assumed to be in *productive*[2] use in some manner or form in an on-going dynamic equilibrium; human society does not go through the effort to design, construct and maintain buildings for no reason. This ratio is a statement about the built environment needs of a modern society and modern economy: for whatever ergonomic (or is that economic?) reasons, our society needs around six times the floorspace for people to live in, grow up in, sleep in, wash in, cook in and rest and play on our own (or in the family) in as we need collectively to do those other principle activities (shop, work, socialise, play together in larger groups and smaller scale storage).

Being an indicator about the floorspace requirements of a successful free-market economy, then logically this ratio also holds for any large metropolitan area, say the whole of Birmingham and surrounding towns and cities. Reducing in area size, it probably continues to hold for any major individual city, say Bristol. The built form of such major city must to a large degree satisfy the economic and social needs of its inhabitant population; if at any point in time the physical does not support the needs of society or economy, then by deduction the free-market will help to re-assert the equilibrium. But at some point, homing in on smaller and smaller areas, this average ratio clearly breaks down: for example, a quiet residential neighbourhood does not include all the workspace and the shops needed by the population of that neighbourhood.

2 At the margins, there is redundancy in all the land uses, which tends to even out over time (conversion of offices to residential, etc.) through different economic cycles. The ratio still holds.

Box 5.1: National Floorspace Ratios

According to the English Housing Survey of 2010–11 (National Statistics-1, 2010) there are 22.4m households in England and the average house size (including flats and houses) is 92 sqm. This gives a total built area of 2,060m sqm of housing in the country. According to the Commercial and Industrial Floorspace and Rateable Value Statistics (National Statistics-2, 2012), the total sum of retail, all types of commercial space and warehousing is about 350m sqm across England. This sum excludes factories. These numbers give a ratio of retail, commercial and warehousing to residential floorspace of around 1:6. Both sets of numbers include redundant space; for the purposes here of identifying an approximate ratio, such redundant space will not significantly alter the conclusion.

The number taken for retail, commercial and warehousing includes all those types of land uses that could reasonably be mixed with residential land use and be located within urban areas without causing nuisance to local residents. That said, some of the warehousing now in use could not reasonably be accommodated within towns (mixed in with other uses), suggesting that for the purposes here the 350m sqm is a slight over-estimate. Factories have been explicitly excluded as not being appropriate for mixing with other uses; they are normally (nowadays) located on the edge of or out of town or in separated industrial areas, together with some of the modern larger-scale warehousing.

This calculation manifestly omits various other land uses in a town, such as churches, temples, mosques, schools, major hospitals, sports arena and so on. These other uses have not been included in this calculation for the following reasons. Firstly, such other uses are not readily mixed vertically with the main land uses of retail, commercial and residential; such other land uses tend to stand alone within the urban landscape, regardless of the density. Secondly, as discussed in Chapter 1, such other land uses do not experience quite the same accessibility drives as those of the bulk land uses (retail, business and residential). Movement patterns to and from these other land uses are more variable and so the locational relationship with the main land uses is less fixed by economic and functional requirements and more likely to have arisen as a consequence of the historical narrative of a city.

The Compact City

At the turn of the Millennium, there was much discussion in the urban planning field around the notion of the compact city. The debate, in large part, arose from comparisons between typical urban form in the UK and US (tending towards sprawl) and that seen in many European cities (being generally more compact). The European compact city model, where average building heights are around six to eight stories, represents a good place to start in considering the relationships between different land uses and the consequences for town structure. The compact city will be revisited in Chapter 11 in consideration of energy consumption of different urban forms.

Consider a theoretical compact city, perhaps the size of Bristol. If the ratio of floorspace, introduced above, were stacked vertically, the rather obvious result would be an average of six floors of residential over a ground floor of other uses across the whole urban area. This typifies the concept of the compact city; it represents a relatively very dense urban area, somewhere in the region of 250 dwellings per hectare (detailed discussion on measuring and understanding urban density is provided in Chapter 6). There are very few locations in the UK that approach this urban density over any significant area; even most of Mayfair and Marylebone in central London are not built to this density over any significant area. To observe cities with sizeable parts of them built to this level of urban density one would have to visit central Paris or other European major cities or parts of Manhattan.

Now imagine, momentarily, living in such a high-density location, where all buildings across the town were on average around seven or so stories in height and comprised a layer of non-residential use at ground level. Drawing on experience of the denser parts of London, it would be a reasonable extrapolation to suggest that anyone living in an area of a city built to such a density could easily access most (if not all) of their everyday needs within just a short walking distance from home. They may need on occasions to travel further, probably by public transport; but, for the most part, their work, all their convenience shopping (and probably all their comparative shopping too) and their amenity needs (sport, leisure and socialising) could be done locally, within a walking or cycling distance of their home.

Looking objectively at the urban tableau, and not just from one individual's perspective, within each area of town, no larger than a short walking distance in length and breadth, there would be provided all the diversity that is seen within any city: pretty much every different land use (a wide variety of shops and eateries and drinking establishments, many kinds of businesses, various leisure opportunities (theatre, cinemas, parks for outdoor amenity), a selection of places of worship and a host of other facilities, even potentially some commercial storage space).

In that the 6:1 ratio represents an economic indicator of the floorspace needs of a successful society and its economy, the dynamic between residential and other uses works both ways. Whilst a person living in such a high density area would be able to access by foot all the services and amenity, potentially even work places, he or she needs on a day-to-day basis, at the same time there is sufficient local population living in an area, above the commercial uses, to provide the needed workforce and customers for the economic and social needs of all those other non-residential land uses. In retail terminology, there would be sufficient catchment spend within walking distance of any retail outlets.

The Super-dense City

There are some places in the world, where parts of cities have been constructed to even higher densities than the compact city. Areas of Hong Kong and Singapore represent the most notable examples of places built to super-densities. Yet, in these locations the 6:1 ratio still approximately holds. In these places, the residential land uses, taking up the majority of the built floorspace, have migrated upwards, moving into the least accessible parts of the city (higher up the elevators). Meanwhile, the retail space must remain in the most accessible and most frequented parts of the urban landscape – on the ground plain, at the base of the buildings, albeit possibly extending upwards one or two floors, to create what are sometimes known in these cities as multi-level Malls or gallerias. At mid-level within the towers, above the shops and below the flats, are squeezed all the other uses, which require good accessibility but cannot afford to be in the most accessible locations – offices, hotels, leisure, schools and community uses.

The result is a sort of layering, dictated by the relative levels of accessibility (determined by trip and dwell-time characteristics) required by each different land use. The residential, instead of lying on the horizontal urban edge is now atop the buildings, in the quietest least accessible part of the city. Each tower, then, is like taking a slice (or transect) out of the natural town and standing it up vertically.

The ground plain of these cities is very fluid in economic terms; the entire ground level is, to all intents and purposes, a permanent market place; shops come and go, but any retail premises is perpetually in demand; when one shop fails the premises is promptly re-occupied by another retail tenant. Travel patterns by the population on the ground are also quite randomised; anyone reaching ground level would have a wealth of choice of shops to visit to purchase their breakfast, lunch or supper, depending on what they felt like that day. But at a statistical level, there is high footfall everywhere, providing a continuous flow of customers to all the shops.

In contrast, the travel patterns between land uses are relatively fixed. Just as a pedestrian resident of a natural town will walk the shortest route from home to the shops and back, so the resident of the super-dense city will necessarily use a specific route to get to the shops and back, generally by means of the elevator in their building. Whilst individual shops come and go, the spatial relationship between land uses remains static. As a consequence the transport infrastructure (mostly vertical in the case of the super-dense city) is fixed in place and remains permanently in demand by the city inhabitants into perpetuity. In turn, these factors maintain on-going stability in the relative land values of different parts of the city; the ground floors, being the most accessible, are automatically the most valuable.

As will be seen later in this chapter, these constancies in the high density city are not repeated in the low density urban sprawl.

The *Not-quite* Compact City

Where buildings are on average around seven-stories high, the whole ground plain can economically be sustained as non-residential, with all upper floors being homes. Drawing on observation of cities that are actually built to around this density, the ground plain still sees aggregations of different types of economic activity; there is not a uniform mix of shops and offices and different types of shops and offices; there is variation, such that some sections of some streets would be mostly shops, whereas other streets would have mostly office premises fronting onto them at ground level. Furthermore, as noted in the last chapter in relation to the tendency for similar activities to congregate together, across such urban pattern there would be pockets of restaurants, groupings of cafes, a congregation of clothes shops, a street full of estate agents; these represent specialised shopping streets. Within the super-density scenario considered above, such aggregations of similar uses would most likely be expressed within individual tower blocks – the galleria within one tower block specialising in clothes (perhaps) and the retail floorspace within the next tower block being mostly taken by restaurants. And so on. Such granularity is not predictable; it has arisen from the outcome of a narrative of the city – historic coincidences leading to two then four then eight restaurants all seeking to co-locate along a short stretch of street. But once it exists, such pattern tends to remain stable over time.

From the high density, seven-storey scenario, keeping the 6:1 ratio constant, a gradual reduction in density would initially lead to the appearance of mono-use buildings. Flats would come down to street level in certain places. Within the overall grid of streets, there would appear streets that are mostly or purely residential. At the same time, the congregation of similar use types would become more marked along other streets; very distinctively retail and very definitely commercial groups of streets would begin to emerge, where retail streets would be those which (for whatever reason) were more locally accessible. The same ratio of residential to other land uses would still exist across an area measured in several urban blocks in each direction. The ratio is now expressed as horizontal, albeit fairly random, mixing across the urban grain, rather than vertical mixing.

Mayfair in London (mentioned earlier) is a very good example of the *not-quite* compact city. As can readily be seen from a casual meander around such an area, a major proportion of all the buildings are mono-use; vertical mixing of uses is the exception, not the rule. (It is really quite surprising, in the context of the current application of planning policy on mixed-use elsewhere, how few buildings in places as dense as Mayfair contain more than one specific land use.) To further corroborate these observations about distribution of land uses in places such Mayfair, from a development perspective it turns out that the land values generated by offices or flats in areas of this density are broadly equivalent. If a developer were to obtain a plot for redevelopment in such an area (excepting plots facing directly onto the retail streets), then it is essentially a toss of the coin as to whether to build offices or flats. The decision on what to build will be dictated by the current state of the economy, the likelihood and anticipated speed of sales or lets at time of completion and the particular specialism of that developer. Over the longer-term the land values obtainable from either business or residential uses in such locations is broadly the same.

The *Limits of the* Pedestrian City – Heading Towards Suburbia

A gradual further reduction in density from that typical of the very centre of London leads to the next readily observable and economically durable urban form: the high street, which is a generalised shopping street, containing all types of shops. As density falls below that of areas like Mayfair, then the clustering of land uses becomes more and more marked, with the discrete land uses diverging into relative locations according to their different accessibility requirements across the horizontal map. This gives rise to the emergence of the high street – a line of concentration of retail with an immediate backdrop of business uses (see Figure 5.1), surrounded in all directions by purely residential areas. Vertical mixed use very quickly becomes the exception, only found occasionally on high streets themselves and not economically sustainable anywhere else.

At these slightly lower density levels, the optimal configuration of the urban landscape, which can provide maximum accessibility to the shopping from as large an area of housing as possible, gives rise to parallel lines of concentration of retail. In larger cities such as London it is a natural outcome that there be a series of almost parallel high streets running out of town. A good example of such urban form is that of Islington and Camden high streets, running north from the centre of London. These two streets are a significant walking distance apart, but half the distance is quite a comfortable walk from the most distant homes to the shops. So

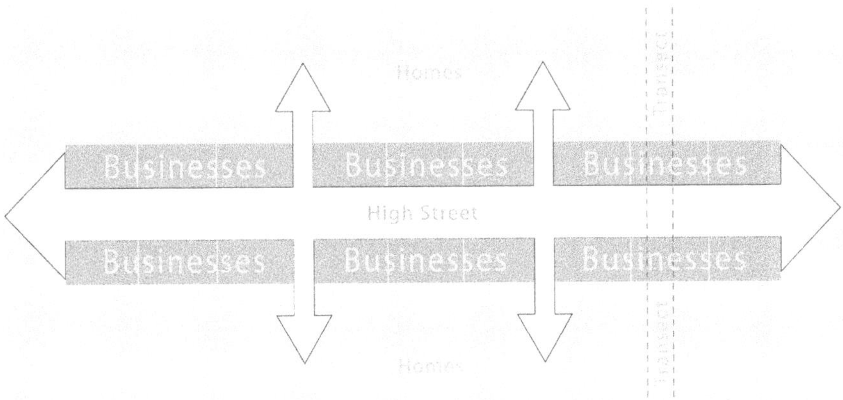

Figure 5.1 The typical high street

all the housing between these two high streets lies within an easy walk of a good supply of shops; and in turn all the shops are easily accessible by foot from the maximum possible number of homes (at around the density to which Camden and Islington are built).

At these densities, the urban ratio between land uses still persists; it is now manifest in a perpendicular transect taken across the high street to the furthest walkable dwellings (about a 10 minute walk from the high street).

At the density levels of the inner London suburbs, such as Camden, Islington and Clapham, there has emerged a clear distinction in land values across the urban landscape. Whereas at the higher densities of Mayfair, the land value of one plot to another is relatively similar, albeit dependent on very local contexts (for example, over-looking a park), at these lower density levels there has emerged a very clear demarcation. Land away from the high street will simply never achieve the values that can be obtained for plots of land on the high street itself; value is entirely dictated by the relative accessibility between one place and another and the immediate clustering of land uses.

Panning outwards from the high street, looking at the bigger picture of the urban landscape, the high street can be seen to represent a centre of activity (see Figure 5.2). Close up, it may be a line on the map; but from a distance, the high street approximates a central point in the urban landscape. In order for the shops to be successful, they need to congregate towards a central point (a centre of retail gravity, so-to-speak – for the reasons discussed in the last chapter). This means that at these moderate densities, the shopping will not be continuous along any road out of a major city; rather it will inevitably be focused at discrete points along a carriageway; this results in the pattern observed in London where those roads out of town are characterised by stretches of street interspersed by stretches of road, imitating in urban vector terms stretches of rail interspersed with stations (as noted in Chapter 3).

Whether the high street is seen as a line through the urban landscape or panning back represents a central locus of activity, there remains a clear correlation between different land uses. Travel patterns between the different land uses remain stable, predictable and perpetuate over time. This in turn defines and sustains different land value contours across the landscape and locks-in the distribution of land uses across the urban landscape – a town centre or high street remains as such into perpetuity.

But when density drops below that which supports healthy high streets, supported by pedestrian access from nearby homes, something interesting and potentially quite problematic takes place.

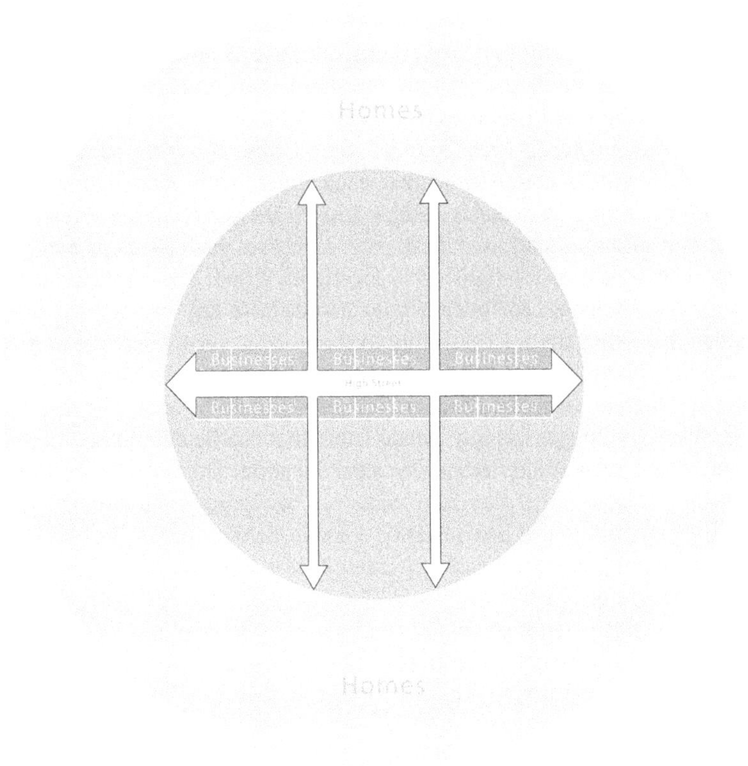

Figure 5.2 High street or town centre?

City Sprawl – Outer Suburbia and Beyond

As will become apparent in the next chapter, the upper-band of what is being referred to here as low density is in fact still a relatively high density – at least by modern UK and US standards. Much of London's outer suburbs are actually built at much lower density. It is only in the very centre of the UK's largest cities that higher densities, as discussed already, can be found. In contrast, in Europe higher densities are typical of many towns and cities for much of their area, regardless of size.

As has already been extensively discussed in Chapter 4, when a significant proportion of a population have access to and begin to use cars, then suddenly they can start to choose where they go to buy their daily/weekly groceries. They no longer need to go to the destination (public or private) that is geographically closest to their homes. Travel patterns between differing land uses suddenly become fluid. In terms of urban dynamics this leads to a dissociation between retail and residential land uses. Homes no longer need to be located within a short walking distance of shops; vice versa, shops no longer need to cluster in a location that is most accessible by foot from a critical mass of housing. When density is allowed to begin to fall below a certain threshold, there emerges as a consequence a disconnection between land uses (see Figure 5.3).

In the last chapter it was deduced that below a certain density *private destinations* begin to compete against each other, with each journey to each destination becoming an exclusive single trip. At the same density threshold, it is apparent that different land uses dissociate; land uses such as shops and housing lose any relativeness or relationship in locational terms. The implication for this is that areas of housing, for instance, no longer have any pre-defined or innate land value that arises from accessibility to shops. Vice versa, there are no singular central nodal points for any area of housing that can be associated with that housing and thereby be natural focal points for retail in the area. The result is that land anywhere no longer has any innate value that can be determined because of its distance and accessibility from any other location. Or another way to look at it, each plot of land either has zero value (or background agricultural value, if no planning permission is provided) or a value defined purely by the land use for which planning consent has been granted. It is irrelevant what land uses are located next-door or nearby.

This is a sudden step-change and can be seen to have arisen as the dominant urban form across America's suburban cities. At these low densities, there is no rationale for any town centre. The very concept of *centre* is meaningless. There

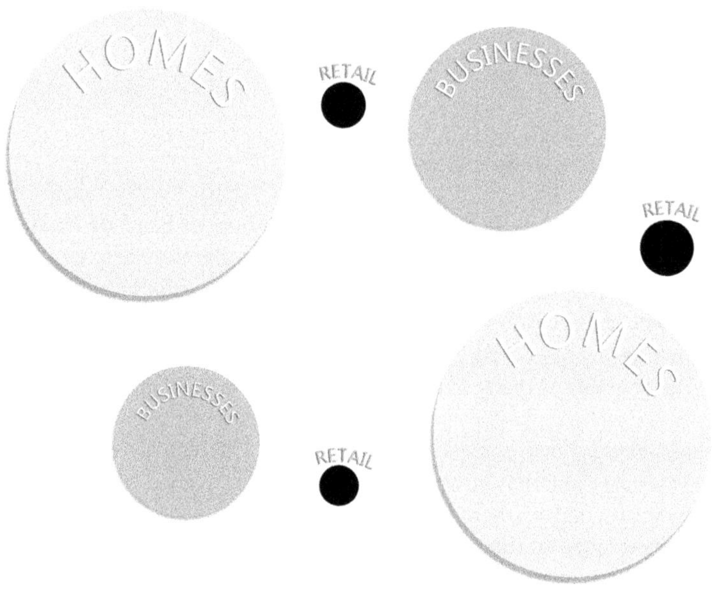

Figure 5.3 Land uses geographically dissociated

is no longer any geographic cooperation between any land uses; everywhere is in innate competition with everywhere else because everywhere is just as accessible (by motorised transport) as everywhere else. There can be no statements such as 'this plot will only ever support housing development' or 'this plot is too valuable for commercial development'. As a consequence any high land values in any particular location are only ever transient, liable to be lost as soon as something better appears nearby, or even further afield. And any investment in buildings and infrastructure is liable to become all too quickly redundant as movement patterns adapt to new destinations on the map. Any masterplanner, who tries to create a natural town structure at these lower densities, is doomed in the longer-term to failure (say, over time spans of 50 plus years or potentially much, much shorter); the urban form simply cannot sustain such structure at these lower densities unless there is an exceptionally rigid planning system to prevent what would otherwise naturally occur, as driven by intrinsic economic and market forces. At these lower densities there is no 'pattern' to the urban grain. Chaos, very literally, rules.

There is, furthermore, no single party that can be blamed; it is everybody's fault. The retailers are simply responding to the desires and choices of the population at large. Every person who chooses to rely on their car for mobility is responsible. But once everyone starts using cars and the landscape responds accordingly, then no one has any choice but to be car dependent. It is an iterative process towards sprawl and chaos.

The timespan, over which change happens, will as discussed in the last chapter vary by land use. In high density situations, land values of and land uses on any particular patch of ground are durable (potentially over hundreds, even thousands, of years), fixed into a patterned system; particular owners and operators change (sometimes rapidly), but the type of land use remains stable. At low densities, however, the landscape will fluctuate over time and any investment in land is both high risk and relatively short-lived. Whilst over a large urban area the ratio of 6:1 between housing and other land uses may still hold, there is no smaller defined area over which one can say 'this area, centred on this town centre, will broadly match the average land use requirements for the inhabiting community'; there is simply no centre to fix on. Any matching of the average requirements (the 6:1 ratio) within any smaller area than the whole metropolitan sprawl will be accidental and unpredictable. The natural fractal structure of the town is lost and has become chaotic.

In terms of the character and structure of the townscape, clearly the low-density urban area is dominated by roads, not streets. In fact, there may be no streets to speak of: after all the shopping often ends up contained within large buildings, Malls, which are (to all intents and purposes) simply *private destination* buildings within the landscape with no relationship to anything immediately around them, other than their own car parking.

It was noted earlier that in high-density contexts it is quite natural for the city to take on a gridiron pattern because of the existence of streets. The streets in the higher density environment are operating not merely as lines, but as spaces with both longitudinal and lateral movement. The existence of lateral movement promotes a perpendicular criss-crossing of other smaller streets. But in a landscape of only roads, where longitudinal movement dominates, then there are no fixes with other aspects of the environment; as considered earlier, a true road has no relationship to the landscape through which it travels. The road can become a wavy line, meandering like a river through the townscape. The wider urban form, as a consequence, can just as easily be gridiron as it can flip to tree-like structures, being the preferred option for some residential developers, or just a random spaghetti mish-mash. The directionality of each road on the map is dictated by the destinations that it links and serves and the topology of the land over which it travels and not by any relationships to other carriageways or any local urban pattern.

In conclusion, at low densities the city very literally loses any structure in either space or time. The consequence is the characteristic urban mat or urban sprawl of the American suburbs. There is no longer any reason for the existence of anything that might constitute a centre of town, a centre of activity for the population or society. There are some serious social consequences to this step change in urban form, which will be explored in more detail in later chapters. But before considering further the social implications of different city forms, the next chapter will provide a more detailed explanation of urban density measurements and the numbers that flow from them.

Chapter 6
Density Thresholds

In the last two chapters, starting from two different perspectives, it has been explained how urban dynamics operate very differently between high and low densities of human activity. High densities tend to lead to stable urban forms, which are held in dynamic stasis because of the collective individual interests of all the different economic operators in a city, working cooperatively for their own personal and organizational gain. Adam Smith's 'each one for himself' market economics actually gives rise to cooperation in the physical world (*Real Space*), even though (or, even, because) economic competition in *Economic Space* (the virtual realm) is so intense. The result is the creation of land value, which becomes fixed in space and endures over time. These are the underlying forces behind agglomeration economics, which when scaled up are the drivers that have created our greatest cities.

In contrast, low densities of economic activity, enabled by a population's reliance on private motorised transport, makes that competition in *Economic Space* manifest in *Real Space*. Rather than land uses coalescing together to create defined urban centres, where all land uses have a relationship to that urban centre, instead land uses dissociate and disperse across the landscape. The result is a townscape where land value can never build up and be stable in any one location over time. Rather all land value is fickle and mobile. The motorcar literally creates a fluid townscape, where any investment in land can only ever generate short-lived returns ... unless, of course, an extremely strict land use planning system is put in place to control urban form, to counter those natural underlying economic forces.

By deduction there must be a density threshold. Below such threshold (and assuming the majority of the population has use of private transport), economic competition becomes expressed in *Real Space* leading to a fragmentation of the town. Again, by deduction such threshold must exist at around the lowest density, at which a pedestrian city can successfully operate. The pedestrian city is here construed to include cities, which make extensive use of public transport.

A Feeling for Urban Densities

Before seeking to make any estimates, set out in Figure 6.1 are the typical densities of various different types of residential urban form. This will help any reader new to urban planning to make sense of what follows.

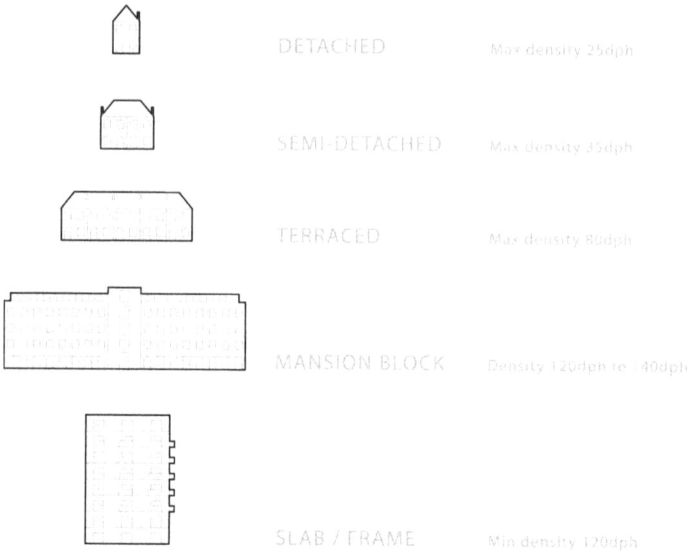

Figure 6.1 Housing typologies

The numbers provided here are all net densities, where the difference between net and gross densities is explained further below.

The Outer Limits of the Pedestrian City

The best way to calculate the lowest density, at which a town will comfortably operate for pedestrians, is to look at statistics for car use. Whilst much work and effort has been expended by governments and many others the world over to reduce car use and to promote other forms of transport (walking, cycling and public transport), the only genuinely and sustainably successful way to ensure that people do not feel the need to resort to private motorised transport is for them to live in urban areas above a certain density threshold. No matter how much cajoling and incentivisation exists to try to persuade people living at lower urban densities to forego their cars, they just don't. Or more precisely, they can't.

So, what urban density do people need to live at to leave the car parked at home or live without a car altogether? Box 6.1 provides the detailed calculations.

Box 6.1: Density Threshold Calculation

Access to Services and Amenities

It is generally accepted that a normal walking speed on the flat is around 4 miles per hour, which equates to around 5 km per hour.

Based on this, in 5 minutes a person will walk around 500m, and in 10 minutes, 1 km. The distance of 500m has been suggested to be the maximum someone will normally walk instead of simply jumping into the car to go and fetch some milk, or equivalent small purchase, or to take the children to school, unless they are intentionally walking/exercising (such as walking the dog). Advocates of sustainable urban design have suggested that to ensure that people are more likely to walk or cycle than use a car, a maximum distance of 250m is more realistic (Thomas, 2003). In the compact city, that equates to a couple of street blocks.

Reading from the inset table (Table 6.1), the number of households required to support a corner shop is considered to range between 2,000 and 5,000. Based on the above distance figures, this means that a population of at least 2,000 should be located within 250m of each corner shop. These numbers suggest a very minimum gross density (see Gross vs Net in Box 6.2) requirement of approximately *50 dwellings per hectare* to sustain corner shops at a sufficient geographical frequency to provide a walkable city environment. Similar figures (see Table 6.1) can be seen to apply to the primary school and the doctor's surgery.

Table 6.1 Creating a Walkable Neighbourhood

Centre Type	Distance	Facility	Population
Local Hubs	150–250m	Primary School	2,500–4,000
		Doctor	2,500–3,000
		Corner Shop	2,000–5,000
Neighbourhood	400–600m	Community Ctr	7,000–15,000
		Pub	5,000–7,000
		Post Office	5,000–10,000
District/Town	2–6 km	Sports Ctr	25,000–40,000
		District Shops	25,000–40,000
		Library	12,000–30,000
		Health Ctr	9,000–12,000

Source: Thomas, 2003

Car Journeys to Access Amenities

The average length of all car journeys made is around 7 miles. Given that a proportion of journeys are very long, this means that to achieve this average a very large proportion of journeys are very short.

UK national statistics (2013) infer that 67 per cent of all car and van trips made by the whole population were less than 5 miles in length. 18 per cent of all car and van trips are less than 1 miles in length and over 20 per cent of trips under 1 mile in length are made by car. Furthermore 64 per cent of all trips made are by car, with walking accounting for only 22 per cent. Furthermore, for journeys between 1 and 2 miles in length, 58 per cent are made by car. In the US the figures are even more stark: around *two-thirds of all trips under 1 mile are made by car* (Federal Highway Administration, 2001).

These figures all point to the fact that people will be lazy (or that they are busy and time-pressured) and that distances to facilities and amenities do indeed have to be very short to be certain that people will walk or cycle rather than use the car. This corroborates the observations above regarding walking distances.

The magic number that falls out of the calculations in Box 6.1 is around 50 dwellings per hectare. This number of 50 dwellings per hectare must be a gross number. This simply means that no matter what other land uses sit within the landscape (a pocket park, public gardens, a church, a doctor's surgery, a community centre, etc.), there must still be 50 dwellings per hectare. This means that in net terms (i.e. excluding all the other land uses) any purely residential parts of the townscape would be at a higher density. In practice this requires an urban form of mostly terraced housing such as that found in London's inner suburbs, such as Islington, Camden, Clapham, Hammersmith and Fulham. Given the imprecision in this calculation and that based on the deductions in Box 6.1 the figure of 50 dwellings per hectare is an absolute minimum, it might therefore be better to provide this minimum as a range: below 50 dwellings per hectare cars are absolutely essential, whereas above 80 dwellings per hectare cars become optional.

This estimate of 50 dwellings per hectare is supported from other indices[1]: at what density does public transport become viable? According to research by Newman and Kenworthy (2006), supported by a raft of other literature (including that done earlier by Pushkarev and Zupan (1997)), there seems to be a range between 35 and 50 dwellings per hectare (gross), where public transport begins to become viable. Thirty-five dwellings per hectare may be able to support (just about) an hourly mini-

1 The 50 dwellings per hectare number also appears in a third another area of urban planning, albeit not related to walking distances and use of public transport. Very approximately 50 dwellings per hectare is also the minimum urban density at which district heating networks can operate in terms of both technical functionality and financial viability (Energy Savings Trust, 2008).

bus service – a bare minimum of public transport – where the vaste majority of the population would still be making use of cars. At around 50 dwellings per hectare more significant public transport infrastructure *can* become viable. In practice it is not until a city significantly exceeds the 50 dwellings per hectare that there can be a sufficient frequency of public transport services that a sizeable proportion of the population begin to be willing to leave their cars at home. The 50 to 80 dwellings per hectare range deduced above therefore seems to apply.

Density Calculations

Numerical estimates of urban density are fraught with complications and caveats. Numerous attempts have been made to formulate robust and consistent methodologies for calculating the density of the city. None have been fully successful to-date. Or, rather, no successful metric has been formulated, which can account for the messiness of real city environments. The fundamental problem is that the city is made up of many different components and is fractal[2]. One of the most common aspects, which tend to be misconstrued, are the differences between net and gross densities (see Box 6.2). There is a persistent problematic: what should be included in net measurements (plots, pavements or to centre line of roads) and over what area should a gross measurement be taken? (see Figures 6.1 and 6.2) Given the vagaries around this, there is a tendency for people to measure and quote what they want to tell the story they want to tell (for planning application purposes and the like).

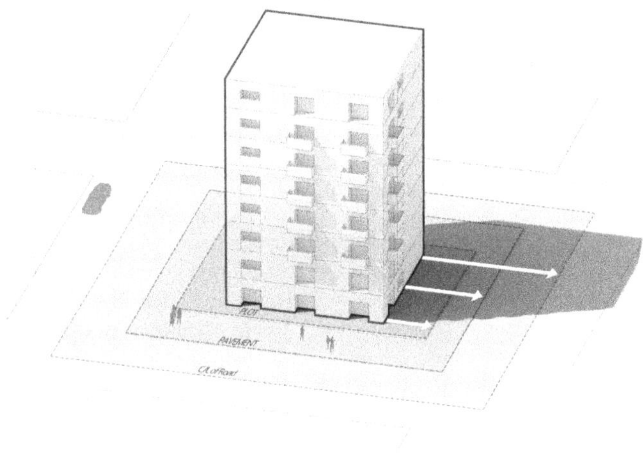

Figure 6.2 Gross verses net density

2 The concept of fractals, as applied to urban form, was introduced in Chapter 3.

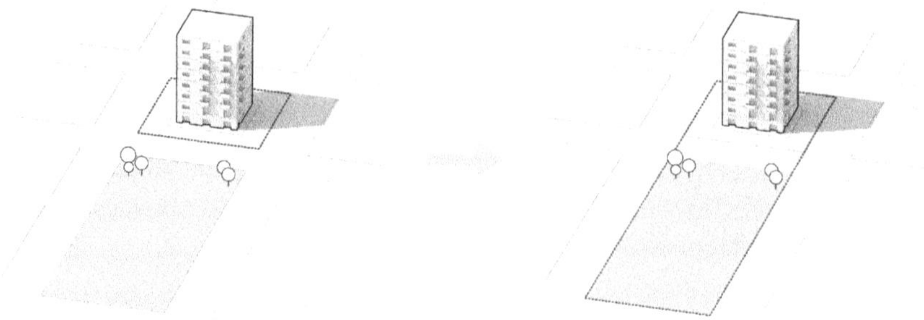

Figure 6.3 Gross verses net density

Box 6.2: Gross verses Net Density

Net residential density measures the area of a site up to the surrounding roads and includes elements that are of immediate benefit to and clearly associated with that housing, such as small pockets of open space. For a small new housing development, it will include any road and car parking which serve only those new houses. In high-density areas, it will tend to include the curtilage of a block of flats and the pavement and half-widths of immediate surrounding roads. According to the London Plan, 'residential density figures should be based on net residential area, which includes internal roads and ancillary open spaces'. For most smaller planning applications, such Net Residential Density will frequently be calculated according to the number of residential buildings to be built within the red-line[3] area of a planning application.

Gross residential density measures a wider residential area and includes parks, schools, the road network passing through the area, other transport services and any other land uses. Larger masterplans will typically include all these other uses and, consequently, the calculated urban densities of such masterplans will normally be closer to the background urban density than planning applications for small developments, where net densities tend to be used.

Care should be taken to ensure appropriate comparison, when considering the Net Density of a new development against a suggested background Gross Density for an urban area. Net Residential Densities will always be higher than Gross Residential Densities. Net Residential Density is quite tightly defined and allows comparison between developments, which comprise only housing; hence its use within the London Plan.

3 The 'red-line area' is a term used in the UK planning context to mean the specific area bounded by a red-line to demarcate a planning application.

The most common measures of urban density are introduced below. Each has benefits and short-comings, some of which are highlighted.

Population Density

This is the number of people living within a hectare or square kilometre. This metric tends to be used more in the world of urban geography than within an urban design or local urban planning context. However, variants to density do become useful in urban design when seeking to estimate the need for certain resources: for example, child density can help to inform potential play space requirements for a new development or within an area of town.

Dwellings per Hectare

This measurement is what it says on the tin. It measures how many households there are within a hectare area. It is the most commonly used, at least within the UK town planning context. It is the primary metric used within the London Density Matrix in the London Plan. It tends to be the most commonly used term with respect to density that is used within design and planning discussions. It is the measurement that professionals seem to understand and relate to best.

However, Dwellings per Hectare can be very misleading. In very central urban areas, where there is a significant proportion of other land uses, then Dwellings per Hectare within the townscape may drop off, compared to more residential areas a little further out. Dwellings per Hectare may also be misleading with regard to local population density, where wealthy areas with lower levels of occupation per dwelling will generate far lower population densities for a given dwelling per hectare number than poorer areas with larger numbers of people living in each home.

Variants on Dwellings per Hectare

The most commonly used variant to Dwellings per Hectare is Habitable Rooms per Hectare. This is intended to give a better idea of actual population density, where number of habitable rooms is more indicative of number of people living in an area than merely counting numbers of households.

Comparison between Dwellings per Hectare and Habitable Rooms per Hectare provides an indication of the number of family dwellings within an area or within a new development: the larger the number of Habitable Rooms above number of Dwellings, the greater the proportion of family homes. This is helpful when seeking to estimate future educational and health requirements of a new proposed community. As noted above, however, consideration needs to be given to the affluence of the area to judge whether habitable rooms will genuinely reflect the number of people living within an area.

Employment Density

Employment Density measurements turn the concept of density on its head. Instead of being a measure of number of people working within an area, employment density metrics are measurements of the average floorspace requirement per full-time equivalent employee of a particular type of land use, such as offices or retail. Standardised employment density measurements (i.e. typically in clothing shops there will be, say, 10 sqm per employee) are then used to help calculate how many full-time jobs a new proposed development will generate (Homes and Communities Agency, 2010).

Plot Ratio (also known as Floor Area Ratio, Floor Space Ratio and Floor Space Index)

Plot Ratio provides a sense of the overall built massing within an urban area. It is a ratio of the total quantum of built floor space within a defined area against the total land area. A three-storey building compared against its own footprint would obviously give a Plot Ratio of 3. But in a similar way to the manner in which Gross Densities of a particular aspect of the urban area (such as the density of dwellings) are rapidly diluted by other factors, so too is Plot Ratio. As soon as roads, car parking and public realm are included within the Plot Ratio, when considering a wider urban area, then a piece of the city with three-storey buildings would more likely provide a Plot Ratio of around 1 or less. As a consequence, a Plot Ratio of even just 1.5 gives rise to, what would feel like, a relatively high-density urban area, equating to 4-storey buildings with relatively narrow streets. Higher Plot Ratios of 2 and 3 can be encountered in central urban areas with taller commercial office blocks, which have very large floor plates. Because residential buildings tend to have smaller floor plates, purely residential neighbourhoods will rarely provide Plot Ratios above 2, even in very dense locations.

Plot Ratio is used in some contexts in the US and Australia as a maximum requirement to limit the amount of new development in an area – to keep density low.

Where Next?

Armed with these rules of thumb with regard to urban density, the next two chapters will make much more sense. In the earlier chapters of this book, consideration has been given to the macro-structure of towns and cities. In the next two chapters the focus will move to the micro-scale structure, in particular of residential areas. In Chapter 10, the macro and micro will be brought together to attempt to provide a combined understanding of the set of drivers, which influence urban dynamics and ultimately urban form.

Chapter 7

A Journey into Town – Part 2

The first half of this book has focussed on the network of roads and streets that make up the transport infrastructure of any town and city. An explanation has been provided as to how towns and cities function from the perspective of movement patterns, arising from all the members of the population each going about their daily routines to live, work and play, including providing services to each other. The macroscopic forms of towns arise out of this day-to-day life; this will be revisited in Chapter 10.

Out of this analysis it has become apparent that, in travelling from the countryside into the natural historic town or city, the character of the thoroughfares gradually transform from being transportation routes, whose only raison d'etre is to carry humans and their goods from one destination to another, towards something quite different: the street. Streets may still be made from tarmacadam, but they represent public space, which has purpose other than enabling people to get from A to B. The essential element of any street is its sides, enabling access onto and off the public highway and facilitating human interaction along its edges. This engenders a situation where on the well-functioning street there is more human activity transverse to the line of the street than along it. Through this, the street becomes a *place*, a node of human activity, a place of intersection of movement patterns, a *place* with a natural high dwell-time in a public environment.

Once the concept of urban vectors has been introduced as a way to help appreciate the underlying urban dynamics, then it becomes apparent that the perfect road does not constitute space at all; it is effectively a single dimensional object. At the other end of the spectrum, the perfect street is very much two dimensional, representing genuine space on the urban map. And where streets intersect, then incidental destinations are created within the city fabric, which have been termed in earlier chapters as *public destinations*. These locations become genuine *places*.

True roads and true streets represent ends of a spectrum of carriageway behaviour (with fractal dimensions ranging from 1 to 2), which influence the character of the urban landscape, determining in large part the efficiency of land use and density of development around them. Roads naturally promote lower densities, pushing activity away from their sides: streets, the opposite.

In this and the next chapter, a second journey into town will be taken. This time, instead of looking at how the roads convert into streets, we will look at how urban form naturally adapts from out-of-town to town centre. Some of the observations, which are made along the way, may seem superficially rather obvious; but considered from the perspective of how the public and private realms of human activity transform along the journey and why, such observations provide insight into how designers should think about what is appropriate at different intensities of human activity. Given, as already noted in Chapter 5, that housing forms the major part of our built floorspace, the focus to start with is primarily on the residential; the derived principles, however, relate to all built form.

To re-emphasise, the analysis that follows pertains to urban environments that would occur organically or naturally. There were many experiments tried out over the twentieth century, where ideological approaches sought to challenge the natural norms and provide alternative urban structures for people to inhabit. A sizeable proportion of these ideologically driven designs proved erroneous and many are now being torn down – witness the numerous 1960s and 1970s estates across London, which have been or are being demolished. The common mistakes, which run across all these different approaches, arose from a failure to appreciate what naturally works in terms of urban design (in particular the demarcation between public and private realms) at different intensities of human habitation and activity.

From Countryside to Outer Suburb

In the rural context, buildings sit within space. As a generalisation (recognising exceptions such as farmyards, where external enclosure of space may be required for the containment of livestock), the only space fully enclosed is the indoor space providing a protected, homeostatic, comfortable environment to sleep, cook, eat and play (more of which in Chapter 8). Rural dwellings tend to sit within a precisely defined space – the garden or farm – where such land represents the private external amenity or dedicated working space associated with the house. Such amenity space may include other buildings (the garage, the shed, the swimming pool, the home office), all of which provide additional specific functionality and amenity for the sole benefit of the occupants of the dwelling. In urban design terms, the key factor here is that buildings sit within space, specifically privatised space associated with the dwelling.

With the obvious exception of national parks,[1] in the country there is no public space. There are no streets as such in the countryside. The roads, which criss-cross the rural landscape, are, as previously noted, single dimensional objects; they may physically occupy space, but the use and purpose of that space is for people passing through the area. Apart from allowing access to and from a dwelling, such space does not provide any amenity to the local inhabitants. Any amenity that they have sits within their own private plot.

Travelling towards and into the outskirts of the idealised town, initially the only observable change is that dwellings come closer together. To draw on Von Thunen's model, historically agricultural activity nearer to the town was focussed

1 Country parks and the like may be situated in the countryside, but they are very definitely not a product of the countryside.

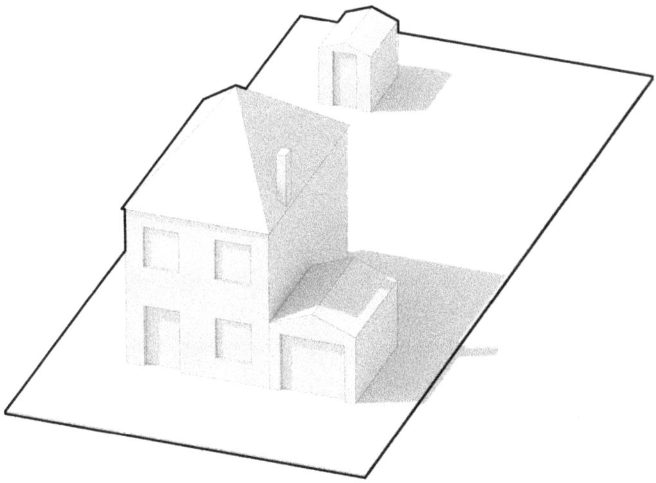

Figure 7.1 Detached house in garden

on goods, which required less land per unit value of produce, so farms and farmsteads near the town would tend to be smaller than those further away. In the modern context, we simply observe an increasing density of detached houses, each situated within its own plot of private land. There are still no streets, so there is no public space; all land is essentially private. Moving into the outer suburbs, little changes except houses become packed ever closer together. As packing increases, gardens become ever smaller; each house forgoes private external amenity space in exchange for greater proximity to the town centre.

As the carriageways are all roads, there is no structure to the town yet.

The first point that any obvious change in urban form can be observed is the appearance of semi-detached houses. For houses with a reasonable rear-garden, then this tends to happen at around 20 dwellings per hectare (note – well below the density required for public transport viability). As density is increased, this transition to semi-detached is the first time that there is any compromise in the buildings in order to retain usable external amenity space. As detached houses are packed closer together, then the space between them becomes less and less useful; the semi-detached solution collapses the space on one side so as to create more usable space on the other. But apart from the fact that two households are now sharing a single building, the semi-detached house still represents a building sitting within the private amenity space, which exists for the sole use of the occupants of that building. In urban form terms, buildings still sit within space. And at 20 or so dwellings per hectare private external amenity (each garden) is still generally large enough that little, if any, public open space is required (a nice to have, but not essential).

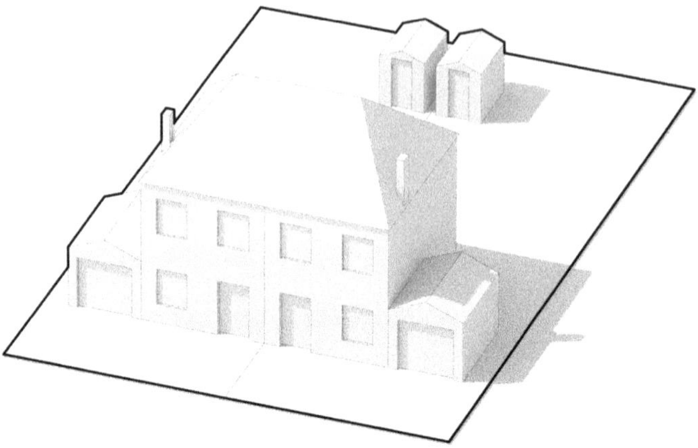

Figure 7.2 Semi-detached houses

Suburban to Urban Transition

Further into town, as urban form is further compressed, the formation of small and then increasingly longer terraces represents a similar design solution to that which created the semi-detached house. Houses are incrementally packed together tighter and tighter in an effort to retain a useful amount of external amenity for the occupants. In this respect, retaining a rear garden, patio, courtyard or yard is the priority; space to the side or front is dispensable as being less private and less useful amenity for day-to-day living.

Throughout this process of densification from detached house to terrace, there has not necessarily been any change in house size. The major component of the change has been the gradual loss of private external amenity (garden area) and design changes in terms of configuration and packing of the houses. The homes themselves, at an average level, vary little in size (the size and configuration of the enclosed internal space); there is far greater differential in house size between rich enclaves and poor areas, whether urban or rural, than there is average rural dwelling size to average urban dwelling size: for example, at time of writing (2013/2014), we have the oddity in the UK that the minimum floor areas required in London for all tenures of housing are larger than those required elsewhere in the UK. The Superdensity Report (Design for Homes, 2007), prepared in the last decade by a selection of London housing architects, suggests that on average floor areas of city dwellings should actually be larger than their rural counterparts to compensate for lack of external private amenity. So, as a bit of a generalisation, dwellings themselves remain a pretty standard averaged size from rural hinterland to the very centre of town; but private external amenity clearly does reduce.

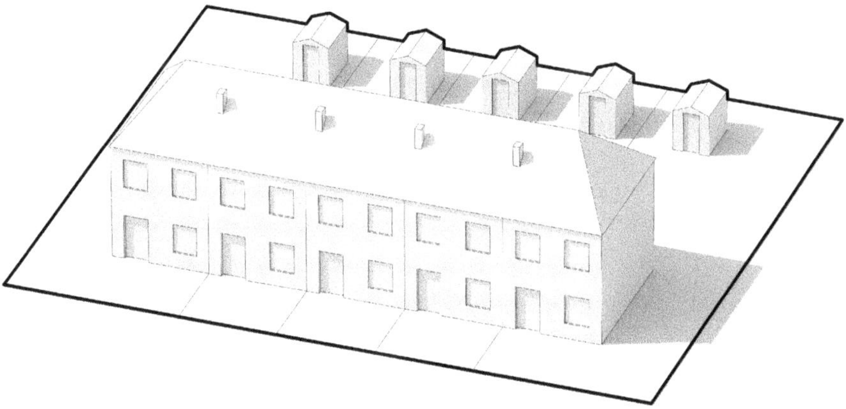

Figure 7.3 Short terrace of houses

As terraces emerge at increasing density, urban form does not initially undergo much change. Semi-detached houses become short terraces of three, four and five households. These are still essentially individual buildings sitting within the private amenity space associated with and dedicated to them, frequently set back a little from the carriageway (see Figure 7.3). But at a certain density, in the region of around 50 dwellings per hectare (that magic number, again), a transition does take place in urban form. It is a subtle but important change and probably demarcates the difference between what we commonly perceive as being suburban and that which most people see as being urban. The change in question represents *a transition from buildings sitting within space to buildings enclosing space.*

When the terrace has become long enough, then it no longer manifests as a building sitting within the private amenity associated with it. Rather it has become a building that, usually together with another terrace backing onto it or four terraces wrapping a square area, encloses space (see Figure 7.4). The private gardens to the rear of individual houses, which by this point of intensification are not much more than courtyards, have collectively become enclosed by the housing to which they belong; side garden area or alleyways have been lost altogether; and front gardens have become compressed into nothing more than thin buffer zones. At this density, most private amenity space has become enclosed by buildings rather than surrounding buildings.

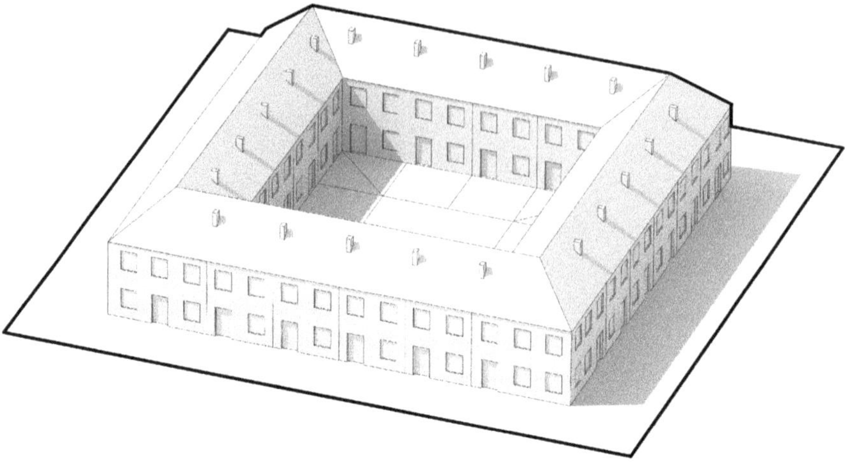

Figure 7.4 Terraces enclosing space

As the front garden shrinks with increasing density, then the character of the urban area as a whole changes. The terraces of housing approach the carriageway such that the road (increasingly street – see below and next chapter) starts to feel less open; it is now bordered by buildings as opposed to open space. At this point any residual front garden may provide a buffer to the public highway, but other than that the space does not provide useful amenity for the occupants.

As homes become packed closely along the road, the side of the carriageway takes on the sense of being increasingly permeable; there are more frequent entrances onto and off the carriageway and given the density there is regular and frequent activity across the door thresholds. At some critical density level (probably somewhere in the range 50 to 80 dwellings per hectare (gross)) (and caveated that the street in question is not part of a rat-run for traffic travelling from further afield) typical movement patterns generated by those living on the road/street come to be as transverse to its line as along its length. Movement by people in the street are as likely to be stepping onto or off the highway to leave or enter a home as along its length, journeying to somewhere else. In making this statement, perception is everything; for those living on one part of the street, people from a distant part moving past are perceived as only journeying along the street. So, if the residential carriageway under consideration is too long, then it may never achieve the feel of being a street because there are always a sizeable number of people travelling past any other part of the street.[2] This suggests a limit to the length of any residential carriageway for it to operate genuinely as a street;

2 This does not mean that the street needs to be a cul-de-sac. Rather, for a residential street to act and feel genuinely like a street there must be more activity across its width (interaction between neighbours) than run along its length.

otherwise it remains towards the road end of the road/street spectrum – with the consequences for the residential community as discussed in Chapter 2.

In the well-designed town, with residential carriageways that are not too long, when housing has reached a density equating to closely packed terraces, then the roads have gained the character of streets, as defined by the movement on them. This transition from road to street is an important urban design change, which significantly benefits the local residents. If the road has genuinely become a street, then by definition (according to the logic described in the earlier chapters) the carriageway has become a space in the city – it has become two-dimensional (or perhaps more two-dimensional than one-dimensional). In becoming a space, the street confers back to the residents amenity in the form of public space, which begins to compensate for the loss of private amenity. The emergence of streets further suggests that different property ownerships are beginning to cooperate on the landscape thereby generating urban cohesion and mutually supporting land values.

In practical terms, in becoming an amenity space, the carriageway can become used by the local residents for purposes other than just driving along it: for example, street parties can easily take place, children can begin to appropriate this domain and use it to play on. Private gardens of any usable size (at least from a child's perspective) may have been lost, but a larger shared, public space has been gained in their stead.

The Urbs

To be a street, a carriageway does not need to be a high street with shops; it simply proved easier to explain the nature of the street, and how it contrasts to the road, in a retail context. Carriageways of any type increasingly operate as streets within the town wherever density and intensity of human activity is high enough that a sizeable proportion of the human interaction is greater in a perpendicular direction to the line of the carriageway.

The first appearance of streets, as you journey into town, is absolutely critical in urban design terms. According to the above line of thinking, the emergence of *the urban*, in contrast to *the suburban*, has two major elements to it – one which is very visible and one which is somewhat more difficult to pinpoint.

The urban part of the city, by most people's standards, is reached at that point when buildings no longer sit within private space, but rather buildings visibly enclose private space and the public highway and public spaces generally become bordered continuously by buildings (see Figure 7.5). Where there are breaks in the buildings, then these are generally for streets to access areas beyond rather than providing private space between private properties.

The less visible change is this natural emergence of public space. This is not artificial interventions to create a larger park or children's playground; rather it is the point at which roads are no longer just roads, but provide a degree of public space too, which is of essential benefit to the local inhabitants; furthermore, at this point of intensification, streets can begin to intersect to create *places* (nodes on the map of the town). And the formation of streets, because of the inherent lateral movement patterns that they promote, begins to give the town some proper structure – the emergence of the grid.

In the modern UK the notion of the quiet residential street as public space has become quite difficult to find or even imagine. The car has had such a major impact on our lives that even where roads would naturally operate as streets (or have historically been streets), then they have frequently become designed by the traffic engineers to deter any use other than driving along or parking on; children's play and car movements are generally considered to be mutually incompatible. But think back only a few decades and carriageways in areas of this level of density (tightly packed terraced housing) would frequently have been a hive of activity and social interaction, especially providing good public space for the younger generations to play (even ball games permitted). Given that the private courtyards to the rear of the houses did not provide much opportunity to release youthful energy, the street as public space provided an important compensation for such higher density living. These were real residential streets and the social activity that they exhibited suggested a high level of social cohesion. Such active residential streets characterise the types of urban form aspired to by all those who promote shared surfaces and home zones[3] within new modern residential developments.

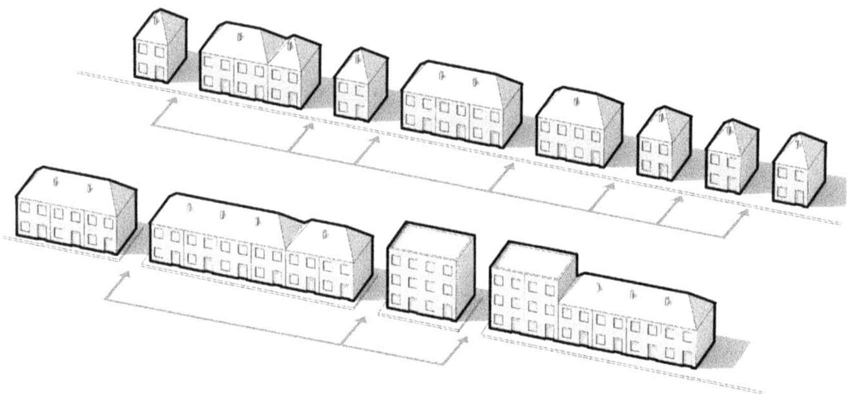

Figure 7.5 Urban verses suburban roads/streets – the visible change

3 Shared surfaces and home zones are the names given to various initiatives that have been promoted to try to reclaim residential streets as public spaces and not just designed as and then treated as roads for cars.

At the point of the tightest possible packing of houses, a threshold has been reached, where, from a theoretical point of view, individual homes and their associated external amenity space have reached their minimum acceptable level. Proceeding inwards into town from this point, all changes in urban form should ideally have to provide occupants with the same size of dwellings and a similar quantum of external private amenity. In practice (at least in the UK of late) this does not happen; private external amenity is often sacrificed yet further to the point of miniature balconies or no balconies at all. For the built form of an urban area to function adequately for the inhabitant community, there must, regardless, be some compensation for loss of external private amenity through the creation of more public open space, whether by means of carriageways becoming more and more street-like or through the creation of incidental public spaces, such as public squares, piazzas and parks.

Low Rise to High Rise

For the traveller into town, so far all development has generally been no more than two (occasionally three or four) stories. Critically all property ownership has directly mapped onto a dedicated area of surface space on Planet Earth; there is no sharing of surface land for private purposes; the only shared space is the emergence of streets providing public space, which is accessible to all and owned publicly (with the clear exception of Gated Communities – see Box 7.1). But in design terms a limit is reached, whereby residential packing can no longer be increased with houses alone.[4] It becomes necessary to stack. The outcome is the appearance of both flats and the occasional (but not frequent) vertical mixed-use building (see Chapters 4 and 5).

Box 7.1: Gated Communities verses Communal Gardens

Gated Communities are those locations where a small suburban residential enclave has been gated off from the rest of the city. It will usually comprise detached or semi-detached housing – rarely anything more dense. As an urban form it is something that can elicit a strong emotional response from urban design professionals, usually in the negative. The simple reason for this is that it involves gating off or privatising that which should (or would otherwise) be public space; it represents the privatisation of a street; the Gated Community is thereby expropriating a significant extra area of space for private use.

4 This limit is reached by a combination of factors – space for car parking, private amenity space falling to a level at which it is no longer functionally useful and the packing of houses into terraces. In the latter respect, houses become thinner and deeper until a point at which they are no longer functionally acceptable.

In contrast the Communal Garden represents the merging of private land to create a private communal domain. It tends to occupy land at the back of buildings which could not otherwise be used for public benefit and would otherwise have been private; in fact it frequently occupies no additional land at all, in that it might sit as a podium deck on top of car parking or commercial premises at the base of the building.

Generally speaking Gated Communities are created for security purposes. It is a response to a perception (real or imaginary) that the public realm is an insecure and threatening environment; it becomes desirous to create an enclave, which could be considered safe and secure for the residents and often includes active dedicated security. Apart from the obvious statement that this makes about the public environment, the contention felt in the UK seems to arise in large part as a result of envy – that those living within a Gated Community are not only living in large opulent homes, but that they are also turning something which would normally be perceived as public space into their own secluded privatised domain. In a country, where land is precious, this is an understandable emotional response. In contrast, in the US and South Africa, where Gated Communities in the suburbs have become quite normal and where there is much greater space available, then this urban form has become both acceptable and an obvious interpretation of the concept of defensible space at very low densities (see next chapter for further discussion on the concept of defensible space). Furthermore in these countries with high levels of gun ownership, concerns over the safety and security of the public realm may be more real than in Europe and the UK.

The emergence of stacking creates new design challenges. Up to this point, there are two fairly discrete design issues to resolve: the macroscopic urban structure (layout of roads and streets) and the design (interior layout and external appearance) of individual buildings with a decent party-wall; the interior of dwellings can for the most part be designed independently of each other. Once stacking emerges then individual units can no longer be considered in isolation of the surrounding environment – design of the city literally becomes three-dimensional. But all the principles so far ascertained still apply, albeit at smaller scales and vertically as well as horizontally.

Focusing on the ground plain, as density is incrementally increased as the traveller enters the central part of a natural city, then a key observation is that movement patterns are less and less of an inward and outward nature (into and out of town), but rather become multi-directional. More and more roads become streets and to a greater and greater degree become public space for activity rather than corridors for purposeful movement. At the logical limit of this process, experienced in the highest density cities in the world, all the ground plain becomes public space, even the ground floors of buildings. However, at the density to which most of us relate in, say, European cities, which is somewhat short of the extreme of all the ground plain becoming public space, the city achieves a density where all buildings sit within public space.

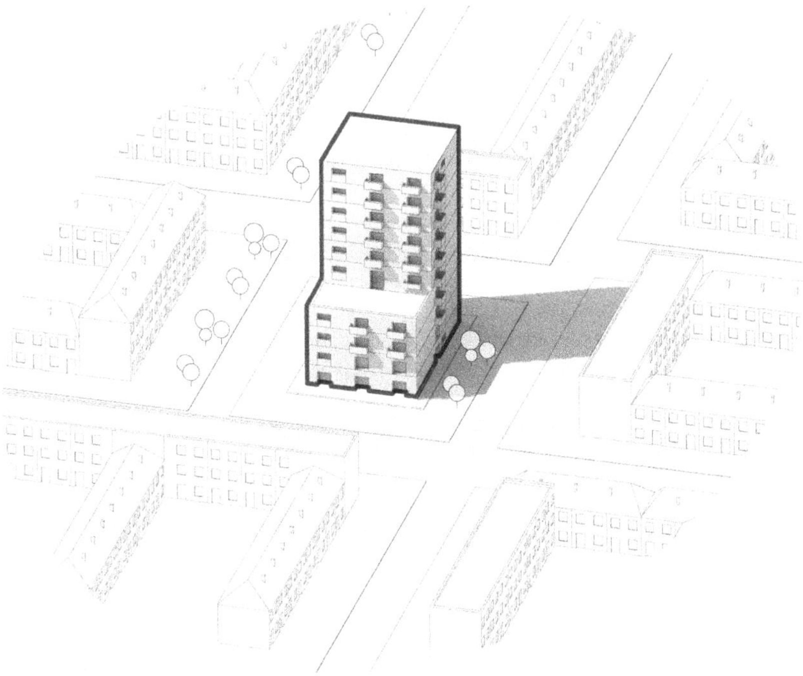

Figure 7.6 Buildings sitting within public space

This progression towards increasing public space at ground level gives rise to another threshold in residential development. At relatively high densities, a good design response for residential development represents the creation of courtyard blocks, where the internal area whether at ground level or at first floor level becomes a communal garden, which is entirely private to the occupants of the surrounding blocks of flats (see Box 7.2 for further discussion on the emergence of communal space). At some threshold, however, it becomes appropriate for this courtyard form to become a quadrangle form where the internal area becomes public space. This only happens at very high densities, where there is sufficient activity at ground level and a high degree of over-looking from the surrounding buildings. It is difficult, however, to judge precisely where this threshold should be: what level of density, and thereby human activity, is adequate to enable a public internal courtyard to be safe and secure through continuous activity and passive surveillance.

So, in the journey from countryside to city centre, the traveller observes a gradual transition from buildings sitting within private space to buildings set within public space. In the centre of town there is no observable external private space; all external private areas have become tightly enclosed by the buildings to which they are associated or set on top of those buildings; all private amenity is thereby cut off from the public world by built structures. Almost all ground level external space is now public and buildings sit within it.

Box 7.2: Legal Demarcations of Space

So far in this description of the urban realm only two types of space have been considered – public space and private space. The definition of these is clear. Public space is space, which is legally accessible to everyone, without exception. Private space, in contrast, is owned (by various legal mechanisms – freehold, leasehold, etc.) and the current owner of such space is able to control access to that space. The legal system for landownership is very much predicated on a two-dimensional land area; the system begins to struggle when the city takes on a 3-dimensional form, as arises from stacking of land uses and ownerships.

When stacking begins to happen, other forms of land ownership also begin to emerge: in particular, communal private land. This represents areas of surface land (or the roofs of buildings) where access is limited to number of people whose only relationship is their link to the property (i.e. a group of residents). The latter point contrasts with employees of an organisation, who are all linked by membership of that organisation rather than some locational linkage, and may by being employees be given controlled access to an office.

Communal space should, in theory, be quite possible at lower densities: for instance, why do not all the occupiers of two back-to-back terraces of houses merge all their back gardens together to create a large communal garden, which could provide a fantastic and sizeable play area for the resident children? It would be quite possible (see Figure 7.4); yet this solution very rarely happens. It is only when flats start to be stacked one on top another that it also becomes culturally acceptable to merge back gardens together to create a communal garden area. It should be recognised that in amenity value terms communal space can never be as good as private external space; as a consequence good residential design nowadays provides each flat with a pocket of private external space (a balcony or equivalent), even where flats open directly onto the communal garden (a little bit of defensible space each).

From Private Rooms to Public Rooms

Over the last 100 or so years, the attitude of professional planners and designers to public space has undergone a dramatic change. Numerous examples abound from the 1960s era of estate building (for example), where there was clearly no thought whatsoever to the spaces between buildings. All design focus lay on the buildings themselves; the public and private spaces between them were a left over after thought.

This attitude to the shape of the urban environment remains pervasive in the housebuilding industry outside our major conurbations. In such lower density contexts, the builder's objective is to knock up multitudes of box homes with handkerchief gardens, as quickly as possible. Regardless of the actual density of final built form, these developers and their design teams seek to emulate the rural housing concept: homes set within private space, where there is no public domain, where highways are all roads whose existence is solely to connect and provide a conduit for vehicles. Attitudes are beginning to change amongst some of the housebuilders, but very slowly.

In London and the other major conurbations, new developments and regeneration projects in higher density areas have turned the approach to shaping the city completely on its head. Instead of designing buildings, say standard house types or (historically) standard concrete slab blocks, and then seeking to cram them together as tightly as possible or to set them out in the most efficient way for construction purposes, it is the structure of the urban landscape and the role and function of the surface space which is considered first (at least, in the better designed developments). The buildings start as grey boxes in the visual impact chapters of the environmental statements, waiting to be designed much later.

By way of example, to help appreciate this changing thinking, imagine moving into a new home. For any individual room, you consider it in the context of the home as a whole and decide first what should be its role and function within the home. Then you begin to decide what furniture to put in and how to structure activity within the room. Sometime later you finally get round to deciding what colour to re-paint the room, what carpet to lay, where to put the lights and what to hang on the walls. This is how we now design our public spaces in higher density areas. We are beginning to treat such spaces as public rooms, where role and function take precedence over looks and aesthetics. The latter are still important; but they necessarily follow once role and function are defined and understood. We now understand, learnt from observing past mistakes, that open space within the city is too valuable at a functional level for the quality of life of the inhabitants to get wrong.

Where the public space is intrinsically valuable, or where its structure is critical to the success of the surrounding private realms (shops, restaurants and other businesses facing onto it), then economic necessity has forced us collectively to design these areas with the right priorities. For example, Malls are very obviously designed structurally and functionally first to create public rooms dedicated to the activity of shopping and then only much further down the design process do the facades of individual shops become considered (like pictures on the wall). In contrast, at very low densities, as already explored, design of the public realm is simply irrelevant; there is no public space to design.

It is, unfortunately, at the middling densities of outer urban and suburban contexts, where we struggle to do things right; masterplanning priorities tend to be the wrong way round; cultural expectations end up over-riding functional need (see Box 7.3). Housebuilders pander to social expectations, such as the desire for detached houses with gardens or provision of private car parking spaces right outside the front door, and pack dwellings together in such densities that every aspect of the urban realm becomes so compromised as to be fundamentally dysfunctional. To appreciate how flawed modern suburban housing estates can become, we turn in the next chapter to look at the essence of the home and the role that it plays for human society. Further consideration will be given in Chapter 12 on the role of town planning for different levels of urban intensity.

Box 7.3: Case Study: Barnet Chalets

The London Borough of Barnet is a suburban borough through-and-through. In many parts of the Borough, it politically struggles with anything construed as genuine urban and seeks to perpetuate its suburban character despite huge population and intensification pressures. Things are beginning to change; but the above summary stands for the last 50 years.

Classic examples, where the urban design has been horrifically flawed and distorted by cultural and political prejudices, are the 'Barnet Chalets'. These arise where developers have obtained groups of adjacent semi-detached houses and sought to replace them with blocks of flats – say four semi-detached homes converted into a block of 20 to 25 flats. The new density is well above the suburban/urban threshold. But instead of designing something urban in character, the solution for which planning is regularly sought and obtained by developers attempts to replicate the suburban concept, albeit at a scale and density which is very definitely not suburban. The result can be described in architectural terms as none other than *semi-detached houses on steroids*. From a distance they look like semi-detached houses; but close up the observer realises that they are several fold larger, frequently being four, even five, stories. These are then completely surrounded by car parking. There is absolutely no amenity space. With the intensified land use, the road outside becomes even more of a road than before.

As isolated examples, these 'semis on steroids' do not seem overly problematic other than exacerbating Barnet's already severe traffic problems. But if these become a precedent and rolled out across larger areas, then the resultant townscape will become entirely dysfunctional for residential purposes.

Chapter 8
Urban Vectors, again

In Chapter 3 the idea of Urban Vectors was introduced to help explain the difference between pure roads and pure streets (recognizing that there is a whole spectrum in between) and the natural forms of junctions created at crossings. The same concept can also be applied to individual plots of land. Before doing so, a quick rehearsal is provided below on what this concept of urban vectors is seeking to portray.

Urban Vectors Embellished

The intention of the urban vector is to make apparent the natural focus of attention (and intention) that people have when they are within a particular environment. When driving along a road, the focus of attention and intention (of the driver at least) is along the highway towards the destination; when meandering along a shopping street, the focus of attention and intention of the pedestrian is on what's on display in the shop windows and which shops to enter. Making explicit these focuses and objectives of the users of particular spaces helps show both how those users of space interpret that space around them and how designers might instinctively design such spaces to meet the collective needs of the users. Furthermore, the way that particular aspects of the urban environment come to be used by others influences how we personally and collectively interpret the built environment: a major highway makes two cities *feel* closer together, yet acts to push activity away from its sides; a successful high street draws activity to its edges causing a natural densification of the built environment around it.

When applying the notion of urban vectors to individual plots of land, the questions that need to be asked are *what is the focus of attention/intention* with regard to, say, a home and *how is such physical object used* by both its occupants and visitors, in particular where are the key points of interaction between people and where does interaction not tend to happen. The easiest starting point in this analysis is the detached home sitting within its garden; from there the analysis will follow the same approach as the last chapter – working into town to consider ever more dense built forms.

Urban Vectors for a Detached House

When it comes to the detached house in its garden, the primary focus of attention would seem to be the house itself. It is the physical object in which the occupants spend most of their time; for most visitors to the property it is the house they are visiting, where they will meet and interact with the occupants – not in the garden. The front door to the house itself is the most obvious key interface point; the front gate to the garden is, in contrast, not. Even where there exists a secure gate to the garden, this is never the end of the journey for the occupant when returning home nor does it tend to be the meeting point between occupant and visitor; the visitor, if welcomed, can expect to be buzzed in and meet the occupant at the front door of the house. The front gate is, therefore, a very transitory zone. The rest of the boundary around the garden is, in similar way to the edge of a road, closer to being a no-go zone; it is far more than just being a hostile environment, as the hard-shoulder of a road; in the case of the boundary line of a property it is intended to repel unwelcome intruders. Inside the property, as a general rule, the further from the dwelling the less well used the garden area is: residents spend most of their time in and immediately around the house and least dwell-time along the boundary line (with the exception perhaps of the dog or a guard employed to patrol that border). For instance, the back of the garden is where the compost heap sits – not generally a place to linger. On the basis of this line of thinking, the urban vectors for the individual plot can be drawn as shown in Figure 8.1.

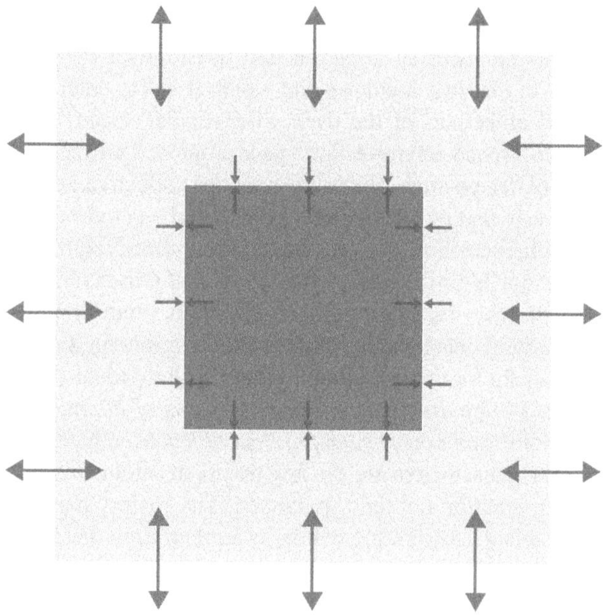

Figure 8.1 Detached house vectors

It comes as no surprise then that, when it comes to designing homes and properties, far more time is spent thinking about the appearance of the house itself and relatively little time and effort is spent worrying about the aesthetics of the border – the garden wall, the hedge or the chain-link fence. The house, then, is the point to which all attention naturally focuses and we design it accordingly. The boundary treatment is a line of avoidance, where least time is spent, and is treated as such in design terms – something to which our backs naturally turn.

Having determined the vectors for the house surrounded by garden, it can readily be seen that this fits comfortably with such property being accessed directly off a road – see Figure 8.2. The vectors along the front boundary line of the house are the same as one would expect for the edge of the road.

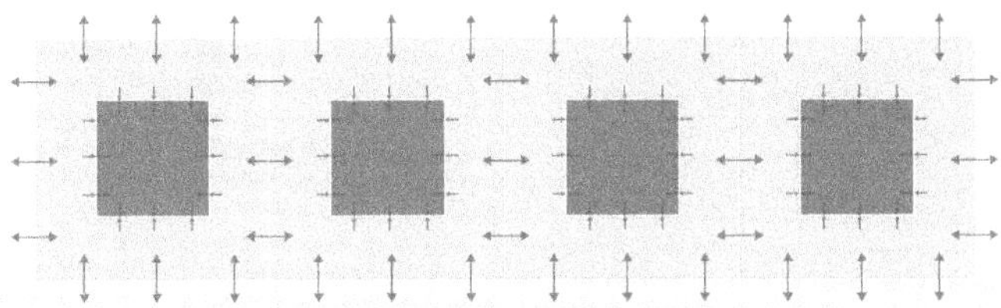

Figure 8.2 Detached houses on a road

Indeed, as can be seen from Figure 8.2 the low-density urban map can quickly be constructed with all the vectors working together. But while the vectors all work together, a question remains: why do house builders desperately seek to avoid producing rows of identical detached homes? There is something admittedly uncomfortable about seeing too many *identikit* detached houses in a straight line (see Figure 8.3 – this doesn't come out in the image, but hopefully the reader understands the point). This will be answered later in this chapter; but first consideration needs to be given on what happens to these vectors as the built environment is intensified.

As discussed in Chapter 7, the semi-detached house is the first step in the process of intensification. Figure 8.4 shows how, as the space between two neighbouring houses is reduced, the urban vectors effectively nullify each other or balance each other out: or, alternatively, each set of vectors could be construed to reduce to zero. From the outside, there is a loss of visibility of that side of the house; it is no longer observed in any way, its appearance becomes irrelevant and no interaction can take place along that edge. From the inside, the same applies as there can be no windows or doors along those walls. The party-wall that is created between the two homes is, therefore, in a sense, a neutral zone, where there are no urban vectors applying in any direction; it becomes a blank façade from both within and without.[1]

1 It would be instructive to learn through empirical research across which borders neighbour disputes in semi-detached homes tend to happen: through the party wall, across

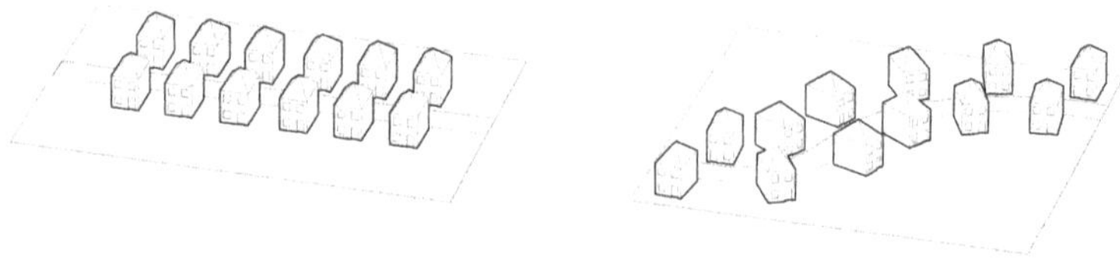

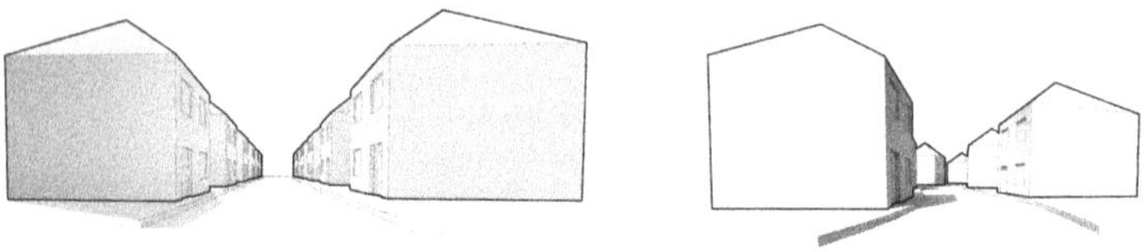

Figure 8.3 Identical detached houses on road or street

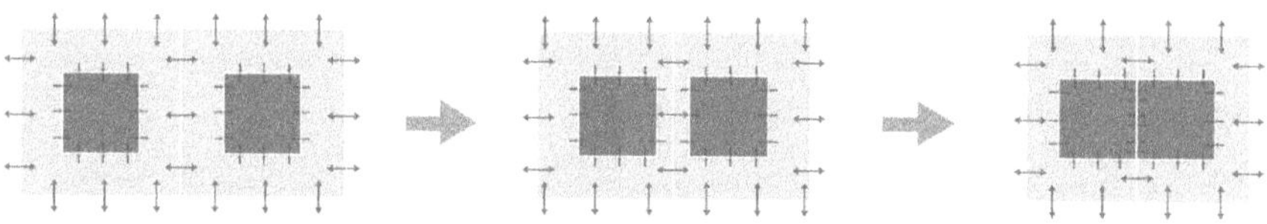

Figure 8.4 Semi-detached and terraced houses – all with extended front gardens

So long as there is no noise transmission through the party-wall, the neighbours on either side can effectively forget that a stranger lives on the other side of the wall. Noise transmission through the party-wall is treated as being a serious sense of invasion of privacy, perhaps in part because it challenges this natural sense of forgetting that anyone is living on the other side of the wall.[2] Inside each house, the area next to the party-wall becomes a zone that neither attracts nor deflects dwell-time. The same principle clearly applies to the formation of terraces. And as with detached houses, a natural street pattern can easily be established with all the vectors working in synergy with each other (see Figure 8.5); the only point where

the fence with the adjoined house or across the fence to the non-adjoined house.

2 Research done by Mulholland Research Consultancy (2003) found, contrary to common belief, that noise nuisance between neighbours is a far more sensitive issue and causes far greater sense of loss of privacy than breaches of visual privacy. Noise nuisance is also far more difficult to deal with and resolve; after all, if a person feels they might be being watched, they can always close the blinds or invest in net-curtains.

it becomes complicated is at corners of urban blocks – a well-known area where urban design often breaks down.[3]

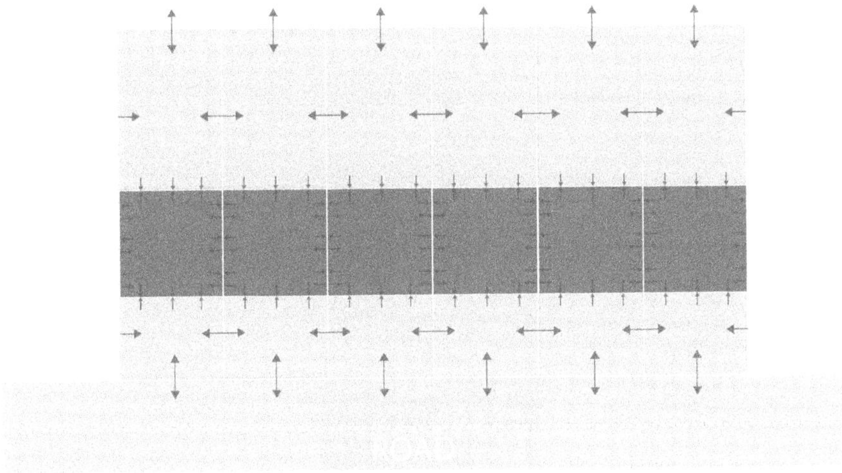

Figure 8.5 Street blocks of terraces

Plot Vectors and Street Vectors

Apart from confirming why so little time is spent considering the appearance of boundary treatments, this analysis has not yet revealed any great insights. But what happens as built form is further intensified and the front gardens of terraced properties reduce as the houses approach the highway? As is shown in Figure 8.6, in this situation we begin to see a conflict between the various vectors. As the boundary line merges into the front of the buildings themselves, it becomes immediately apparent that it is no longer compatible to have a road running directly past the front door (see Figure 8.7).

Using the urban vectors, as a mechanism to help explain the functioning of the urban environment, it can be seen now to reinforce the discussion in Chapter 7 that a section of highway does not need to be a high street (for shopping) to act as a street. The residential street is an equally important component of the urban landscape. Where terraced properties are placed onto roads, an awkward conflict of uses and functions takes place. Similarly, any highway bordered by

3 At a practical level numerous local urban design guides seek to prevent housebuilders from using standard housing types, especially at street corners, where there is a need to address more than one aspect onto the public realm. This is an issue that the Building for Life assessment system in the UK has sought for many years to address – and is finally gaining some traction with housebuilders.

detached houses is rarely going to operate as a street, except perhaps in the case of very short cul-de-sacs (see Figure 8.8) – gated communities, perhaps?

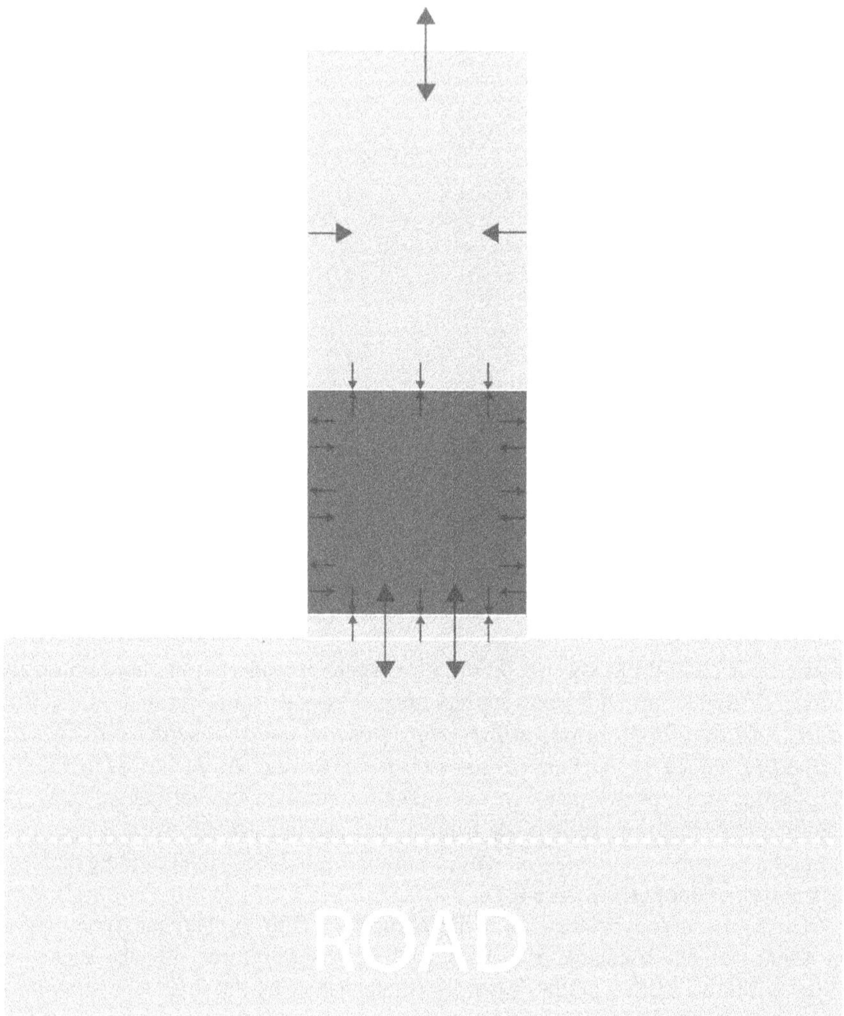

Figure 8.6 Compression of front garden

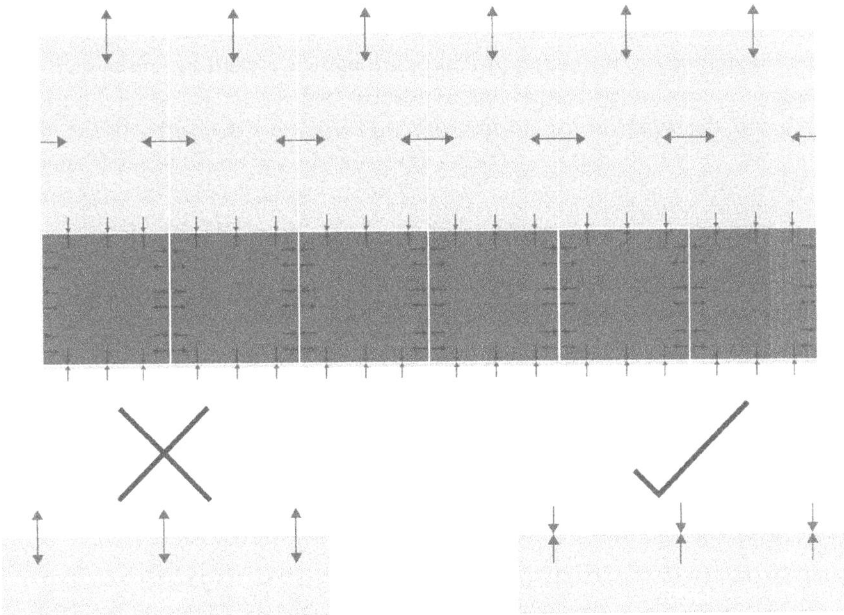

Figure 8.7 Terrace onto a road (cross)/ terrace onto a street (tick)

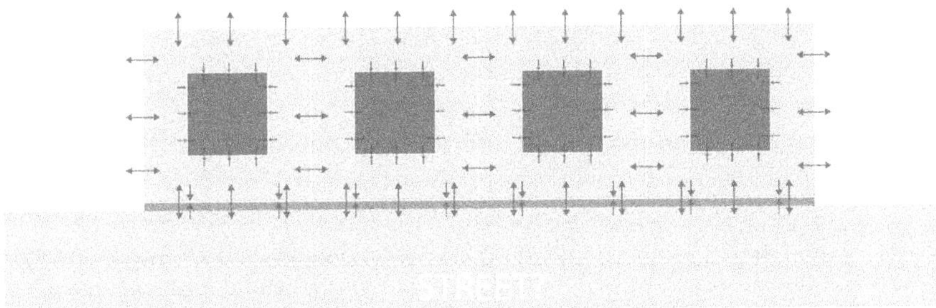

Figure 8.8 Detached homes do not naturally create a street

The potential for conflicting vectors along the front line of terraced residential buildings also helps to explain why this zone proves to be so hard to get right in design terms and why there tends to be such debate over what is the most appropriate design. There would appear to be different camps of designers and design critics on this particular matter. Some, such as the Prince of Wales, appear to believe that the better solution is a more traditional approach where there is no buffer zone between homes and the street, as was the designed solution at Poundbury in Dorchester. This seeks to replicate more historic rural village built forms as well as those streets of workers' houses created by the

Victorians – in both scenarios there were no cars to worry about. The rest or the design community, and probably the majority, tend in this modern car dominated world, to prefer placing a buffer zone between residential building fronts and the street, even up to very high densities and even where there are significant numbers of flats built over ground floor maisonettes. Using the urban vectors to help understand this debate, it is probably technically correct that a residential street will not function to its full effect as a public space unless building faces sit directly onto this space (see Figure 8.9). Where a residual buffer zone is retained, this effectively represents a concession that the space in front of the house remains as much a road as a street and is therefore essentially a transitory space; the buffer zone allows a transition from one set of urban vectors to another.

Using the urban vectors in relation to individual plots can now be seen to reinforce the arguments presented in Chapters 2 through 7 regarding the transition along the spectrum from roads to streets. It is not until the packing of residential properties leads to frequent and regular entrances off the highway that highways sufficiently convert along the spectrum to acting as streets, thereby becoming genuine public space usable by the local community, and not just a through-route for others. That said, if the highway fulfills a connecting or rat-run purpose within the wider road network, then lining such roads with terraces is always going to create an awkward juxtaposition of functions.

While analyzing the interface between buildings and highways, it is worth also considering the consequence of creating blank facades in areas that could or would otherwise function as streets and public spaces. As considered in earlier chapters, streets – to function genuinely as streets – are reliant on interaction between the public realm and the private zones along the edges; streets are places where movement along the highway slows down and pauses and where movement patterns and interactions perpendicular to the length of the highway come to be dominant. If, in creating a new development, a designer chooses to replace a once active frontage with many windows and doorways with a blank, un-interactive façade, it will rather obviously stop or hinder lateral movement. Rather

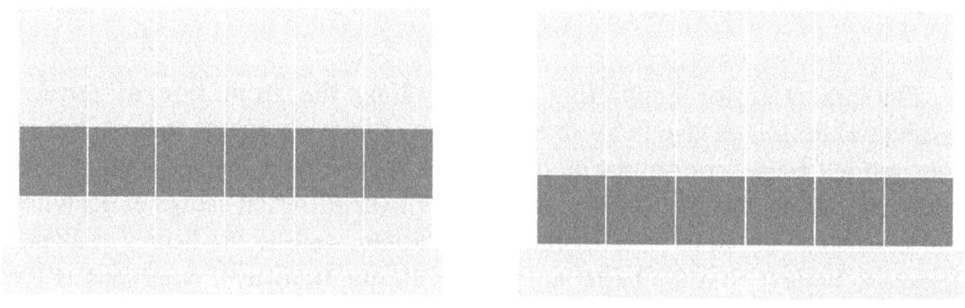

Figure 8.9 Terrace onto street (no buffer) / terrace onto road (buffer)

longitudinal motion and interaction along the highway becomes promoted. All of a sudden, an area that once acted well as a public space becomes more of a transitory road-like place – the antithesis of placemaking. This is precisely what has been happening in the case of developments like the Barnet Chalets considered at the end of the last chapter (Box 7.3).

This analytical approach can elicit better understanding around some of the debates about the implications of different design solutions, which recur on every new project. If an architect has been asked to design a new residential development facing onto a highway that is unarguably a road and there is no opportunity to change how such highway is used, then, for the benefit of the future occupants, the designer has little option but to go with the flow (see Figure 8.10) – step back from the edge, limit entrances facing directly onto the highway and focus on creating a good quality environment on frequent residential streets running perpendicular to the road. Alternatively, if at a strategic masterplanning level there is an option to convert somewhere currently acting as a road into something with more street-like qualities, then an alternative design response bringing residential fronts and entrances right up to the highway edge might be appropriate to help this solution – so long as other mechanisms are put into place to limit and reduce traffic speeds. And where the highway is somewhere in the middle of the spectrum between road and street, the architect can choose a width of buffer zone in front of the building accordingly.

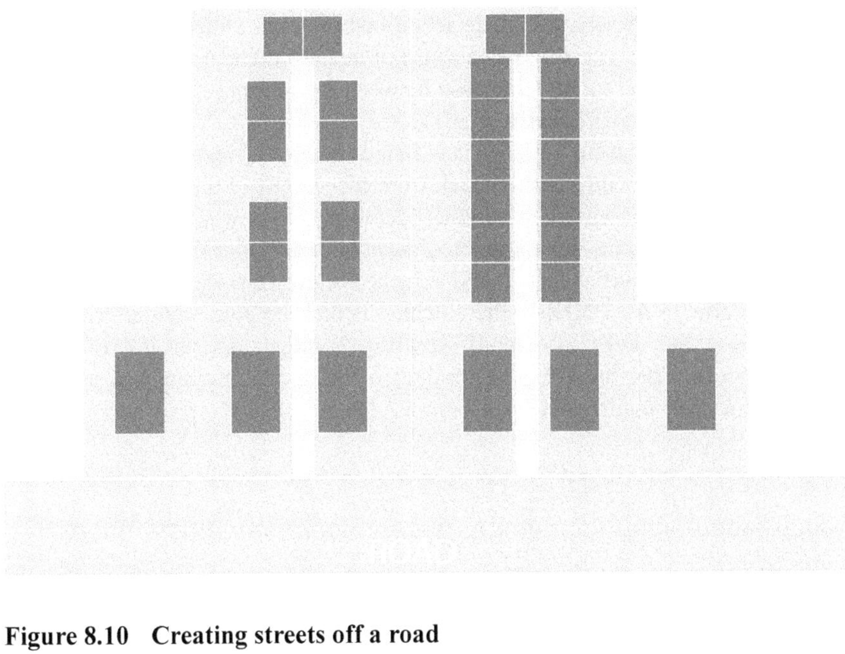

Figure 8.10 Creating streets off a road

Urban Vectors and Appearance

Earlier in this chapter, it was commented how long straight lines of identical detached homes were somehow, somewhat inexplicably, an uncomfortable sight. Housebuilders building identikit boxes (standard housing types) instinctively seek to avoid such design solutions either by creating a variety of house types and peppering these along a straight road or by making the line of the road wavy, curved or somehow higgledy-piggledy (see Figure 8.3). Contrast this to some of the best loved streets, which comprise beautiful terraces of homes, which are all absolutely identical. Why is one urban form embraced and the other eschewed?

Reverting back to the urban vectors, consider for a moment the vectors surrounding a detached property with garden (see again Figure 8.1). Put this observation into the context of the discussion at the end of Chapter 3, looking at how the edges of roads affect the urban form along them, the vectors in both cases are *pushing away*. Consider also the discussion in Chapter 4 regarding urban dynamics and how competition and cooperation are expressed across the urban landscape: at low-densities, all the plots of land are in inherent (albeit latent) competition with each other. The urban vectors around the garden boundary of a detached house make explicit the underlying urban dynamic that all such properties are in reality in dynamic competition with each other (perhaps nobody ever feels that their own garden is quite big enough). As noted in earlier chapters, where competition exists such as in the retail environment, but is contained physically, it becomes expressed through diversification. We may not know this, but we sense it instinctively. In the context of detached homes, such underlying dynamic of competition becomes expressed as a sense of need to diversify their appearance. It is a tension that we intuitively sense between the real physical world and that virtual world of *Economic Space*.

If a housebuilder were building new homes at a very, very low density, such that they were sufficiently pushed back from the street and probably part hidden by trees and undergrowth, then building identical houses would not be an issue. But when detached houses are crammed together at densities that probably ought to require a different urban form (semi-detached or even terraced), as is happening frequently now in the UK, then the outcome is that the housebuilder senses this awkwardness and seeks to diversify his new homes – not just for marketing reasons, but to make the visual appearance of the urban environment somehow more instinctively comfortable (see Figure 8.11).

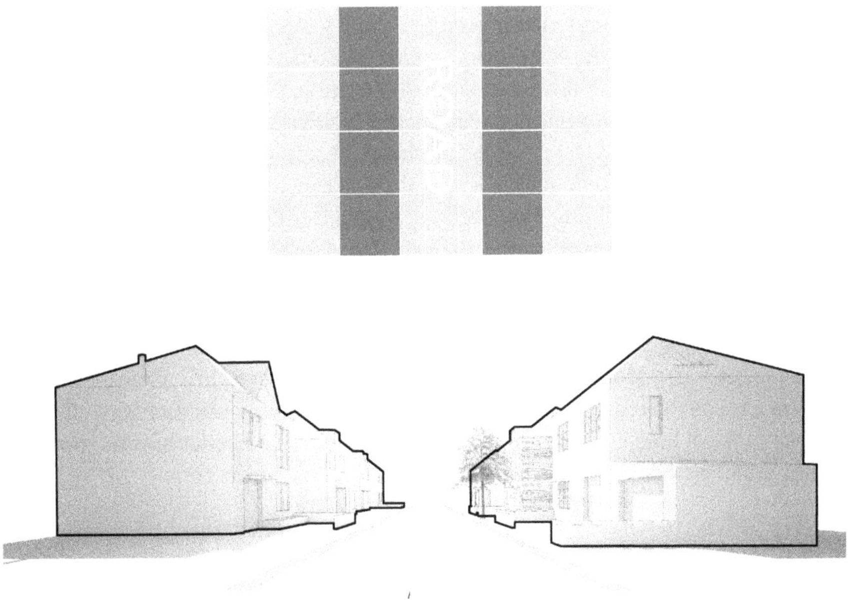

Figure 8.11 Detached properties on straight road need to diversify

As noted in Chapters 4 and 5, at a threshold of around 50 or so dwellings per hectare, competition begins to convert to cooperation across the urban landscape. Competition between different plots of land becomes expressed in different, less physical, ways. In the context of residential development, such diversification at higher densities automatically happens because the precise location of individual properties has inherent differences: for example, from one end of a street to another, homes are closer or further away from essential amenities (where such difference would be meaningless in the low-density context, where everyone is car-bound). Other differences may include variation in tenures along the street. The difference between homes converts from one of appearance to ones of lifestyle and different properties meeting different needs of inhabitants: social and economic differences instead of physical differences (see discussion in Chapter 4 on the concept of Economic Space). Furthermore, as the properties are compressed together, they create singular larger buildings: the short terrace is 'read' in urban design terms by observers as being a single building even though it manifestly comprises several separately owned homes. Being seen as being one building, the component parts are expressly cooperating with each other and it therefore registers better with our senses as a repeating built form.

The other influence on appearance, which may or may not equally contribute to the above puzzle, is the degree to which the design of a building should be considered to be a private matter or a public affair. Where a building sits within its own private domain – in the case of a detached house, surrounded by its garden –

then the design of the house itself is quite arguably a private matter for the owner. Its appearance is of little or no consequence to society at large (see Chapter 12 for further discussion on this) – especially if there is a big, blank boundary wall. However, as the building line approaches the highway, and certainly at the point that the front boundary line has been absorbed into the building line and the building faces directly onto public space, then clearly the appearance and external architecture of the building has become a public affair; at these densities, the building has become the wall and wall decorations for a public room in the city. As intensity of built form increases and as the highway edge converts from multiple private objects to a public phenomenon, then one would reasonably expect the appearance of that edge to change from the private statements of many individuals to a singular expression of something acceptable to wider public society.

Having considered residential development in a purely abstract way, the next chapter looks at housing in the round to seek to understand better how the function of this major part of the urban landscape should influence its form.

Chapter 9

A Place to Live

In earlier chapters consideration was given to elements of the macroscopic structure and natural form of the city, which is driven by all members of the population all seeking to satisfy their daily essential needs – fundamentally to allow trade to thrive and provide townspeople with access to places to buy food and other goods and to participate in economic activity in order to earn a living. In the historic city improved access in the centre gave rise to demand for land, which reflected as higher land value. The outcome was that households near the centre packed together more tightly than on the edge of town; more centrally located urban inhabitants traded off the improved accessibility against living within a smaller area and in closer proximity to neighbours. In larger cities, this has given rise to the sharing of surface land and the stacking of accommodation. All very obvious.

In addition to the macroscopic structure, in Chapters 7 and 8 a description has been provided on how urban form changes as one enters the city in response to the packing together of private land ownerships, the demarcation of public and private realms and the emergence of public space.

All these observations on the macroscopic form do not, however, fully explain the structure of the very local context, in particular of areas that are predominantly residential. To appreciate the influences behind the localised structure of the city requires some thought as to the true nature and role that housing plays for human society. In the same way that the easiest way to conceive streets is through the shopping environment, but the resulting understanding can be seen to apply to other situations (e.g. the residential street), so too the commentary that follows can later be extrapolated from residential to other contexts such as business and institutional buildings and such urban settings.

The Essence of the Home

The quintessence of housing is the *enclosure of space*. This is a bold statement, which can be derived in various ways.

From a purely physical perspective, whilst in temperate climates people may speak in terms of keeping the cold out, it is perhaps more accurate to say the purpose of enclosure of space is to keep the warmth in, to maintain that internal comfortable ambience. But it goes further than merely environmental considerations; enclosure of space is very important in social terms too, from the perspectives of individual, family and community.

A way to conceive better the idealised structure of the home is to consider the classic religious building, the temple, *the house of god*. There is the public space around and in front of the building, open and freely accessible to all; then there is the inner temple, accessible only to worshippers of said god; then there is the inner-inner-temple, accessible only to certain acolytes; then there is the inner-inner-inner-temple, accessible only to high priests; then there is the inner-inner-inner-inner temple, accessible only to … and so on. The further into the temple one travelled, the closer one approached the god in question. The natural physical expression of this is the creation of concentric rings, leading inwards to greater degrees of privacy. This is a very different form of concentric rings to those which Von Thunen deduced arising from accessibility levels in the countryside around a town and which are, in effect, also replicated in the towns structure (as discussed in Chapter 1). In fact, in the residential context, the innermost element is (or should be) by definition the least accessible, as opposed to the most accessible location.

Privacy here is taken to mean the limited number of people allowed (by legal means or other social constraint) to access a space or volume.

The use of the temple as an example is intentional; churches, temples and the like may be communal religious buildings, usually housing a specific god; however, it might be reasonably construed that the structure of such houses of god incorporate an element of idealisation about how a normal home for living should be structured. There exists a remarkable consistency, the world over, in terms of the structural template of religious structures and in turn of domestic buildings. If the use of a temple (by way of example) feels uncomfortable or seems inappropriate, then alternatively think in terms of the structure of the historic castle, which was to all intents and purposes the dwelling for a rather large family – lord or king, his close family, extended family and all their staff and retinue and a local garrison and any other hangers on. Said groupings of buildings were structured to accommodate a social hierarchy, where position in the hierarchy defined what areas someone could access and further created layers of protection from a hostile outside world; privacy in this case also expressed in defensive terms.

In terms of actual physical form of built structures, even in temples, the layers of privacy were and are rarely if ever expressed quite so literally as actual concentric rings. Rather, there has always been a tendency to achieve this outcome figuratively. In the temple context this may have manifest as a series of chambers *leading inwards* (more often than not conceptually inwards rather than physically inwards – the inner-most chamber may simply have been at the back wall of the temple and furthest to get to from the entrance). However, in some religious contexts, the innermost sanctum was genuinely the central location. Regardless of the actual physical expression of a building, the innermost space is by definition the least accessible, most private, physical location – the furthest from the public, outside domain.

Ignoring the religious component of the analogy, it may surprise the reader to realise that pretty much all well-designed[1] housing is structured on precisely the same figurative basis. In the modern secular household, the innermost sanctum tends to be that of the en-suite bathroom. This is the most private space for the occupants of the household, more often than not only occupied by one person at a time to do their most personal preening, healthcare and whatever other private activities. The next most private space is the master bedroom, solely used by or shared by the owner(s) of the household. In a conventional two-floor house, the next level of privacy could be construed to be the entire up-stairs, only generally open for use by those sleeping in the house long-term or invited to stay overnight; but it is not generally public to day-time guests. For many households, the entire downstairs (including the back garden) may be considered to be the next layer of privacy. Historically, the downstairs was frequently sub-divided into front of house, in which to entertain guests, and the remainder into which guests were generally not welcome; this did not just apply to stately homes; go back 100 to 200 years and almost all homes, no matter how small, would seek to set themselves up with a well-heeled, permanently tidy front room and a much lived in kitchen to the rear. For the rural/suburban household, the next layer of publicity is the front garden – the buffer between household and the public realm of the highway. Beyond that there is genuinely public space (unless within a gated community – as described in Chapter 7), freely accessible to all. Clearly it would be an absurdity to seek to design each house as actual concentric circles; but in a conceptual way, whether urban flat or rural farmhouse, in well conceived dwellings this is the effect achieved.

It is next worth considering why houses are designed as such; what function is a home fulfilling by providing different levels of privacy? In its entirety the building, the house, creates a relatively homeostatic environment (as mentioned earlier) to provide relative comfort for the occupants to cook, eat, watch TV, play and sleep and more frequently nowadays again to work. In the modern context it also provides a secure space to stash belongings: possessions that we acquire and deem that we personally need to live our lives. As to why the design of homes has evolved to the creation of discrete rooms for different purposes, the driver could be construed to be in part the very different nature of (and furniture requirements for) different activities such as sleeping and cooking, and in part this issue of levels of privacy. Go back several hundred years (half a millennium or so) and most homes in Britain represented a single hall and not much else; but as we have all become more affluent, gradually the single space for an extended family and servants has subdivided (by means of various trends and blind alleys) into dwellings in their current modern iterated form. In his book, *At Home: A Short History of Private Life*, Bill Bryson (2010) provides a very good history as to how

1 Designed for functionality, as opposed to some architectural concept house designed for appearances or some other effect.

the modern home has evolved from an original single communal hall into a private domain of discrete functional spaces and degrees of privacy.

The functional spaces provide the component parts of the dwelling – the kitchen, the bathroom(s), the bedrooms. It is, however, this notion of degrees of privacy which determines how those elements are put together and structured: who the owner(s) of the modern home might be willing to meet at the front door will be a wider circle of people to those who he/she might invite in to eat and drink with, which is in turn a wider circle of people than those who he/she might choose to invite to stay the night. And so on.

The well-designed home also operates successfully in reverse, enabling a gradual opening up of degrees of publicity. This allows for another critical and fundamental factor when it comes to housing, which must not be underestimated or neglected: the rearing of children. While the home manifestly provides protection from the outside elements and security from outside human beings, it very importantly (*absolutely critically*) provides containment for children, especially young children. As they grow up, the children increasingly become long-term guests in the household, able to use the upper floor but not so readily welcome any more in the master bedroom and its en-suite. When young, the house and the garden provide very important containment – a safe place to play, learn and explore the physical world and social interactions, before setting forth into that rather more dangerous outside world. Over his or her childhood lifetime in the family home, a child very literally progresses outwards through the layers of privacy (from mother's womb) until in adult years he or she becomes just another (albeit hopefully very close) guest in the house, having to respect the layers of privacy no more or less than any other guest might have to. For those very early years, good family housing is defined by its ability to provide safe, layered containment.

It is easily forgotten in a world where the young family is present in each and any household for a relatively small amount of time: but you simply cannot understand the function and thence structure (or at least what should be the right structure) of human housing in its modern iterated form without appreciating its essential role for the rearing of children.

As a general rule, in the UK we have become collectively quite good at designing the internal spaces within homes, especially houses (though they may frequently be a little on the tighter side than those across the water in Europe). A fairly standard structural template has evolved through trial and error, which generally seems to work for most; albeit different cultures have their particular preferences, where some are comfortable (by way of example) with open plan living (for the main living spaces) and others not. Around the world, the structural template used in the UK is pretty standard for all more affluent nations with a tendency towards smaller family units (the preference of the global middle classes).

So, the quintessence of good housing design is the enclosure of space and the creation of degrees of privacy/publicity. And we have collectively become quite good at designing this inside homes. We do not, however, seem to have achieved the same level of sophistication in internal design when it comes to flats. And more unfortunately still, in the modern context we have failed outstandingly to extrapolate this expertise to the urban world outside the front door[2]; instead we have allowed our mobility needs to dominate the wider urban form rather than the needs of our children.

Box 9.1: Degrees of Privacy verses Defensible Space

In Chapter 1 it was commented that no one from the academic or professional worlds has to-date seemingly asked: why are our historic (natural) towns structured the way they are? It is something that is so much part of everyday life that we have come to assume it without question. The same seems to apply for the internal structures of housing. There is, therefore, very little by way of academic literature on which to base the assertion that the essence of good housing is the creation of degrees of privacy. The best direct references found by the author include: *Design for Homes* (2002) and Daneshpour (2012).

There is another area of understanding of good design, which loosely connects to this idea of layers of privacy; it is the concept of *defensible space*.

The idea of defensible space was originally formulated by Oscar Newman in the early 1970s as a result of his research of public housing estates in the US (Newman, 1972). The principles that Newman put together have been developed from their original conception by both Newman and other parties, such as Alice Coleman. In the latter case, Coleman drew on and extended the ideas through her research on high-rise local authority housing estates in the UK (Coleman, 1985). In the practical world of urban design, the thinking of Newman and Coleman has become a mainstay of what is perceived to be good practice. In the UK these have become encapsulated into guidance by the police in the form of Secured by Design (ACPO, 2010; ODPM 2004) and were embedded into CABE's *By Design* guidelines (CABE, 2000).

Newman identified four key factors, which he considered create defensible spaces:

Territoriality – giving people (as individuals and as small communities) clear ownership of the space around their homes (analogous to detached house and garden).

Natural surveillance – ensuring that urban spaces are adequately overlooked, in particular by those who have access to and ownership of those spaces.

Image – the physical attributes of a development that make it defendable (in other words, if a development looks good then people will value it more and seek to defend it). (This aspect of Newman's thinking closely aligns with the more recent broken windows theory originally developed by James Wilson and George Kelling (Wilson and Kelling, 1982).)

2 Some reflections on the history of this are provided in Box 9.1.

Safe zones – making use of buildings and physical structures to define boundaries between public, semi-private and private spaces (essentially the same conclusion as presented in Chapter 7).

Urban designers now use Newman's original concept by thinking of the city in terms of a hierarchy of public spaces, from the most private space for the smallest community up to the public space of high street, a city centre square or city park. This interpretation of Newman's ideas is clearly synonymous with the notion that the successful city provides a spectrum from most intimate private spaces up to most busy public spaces and that the most successful urban landscapes achieve this in a layered way (even if only in figurative or conceptual terms). Where the notion of privacy, introduced in this chapter, extends beyond Newman's ideas is to say that the innards of the home itself and the structure of buildings with multiple homes should also be conceived in the same way – that privacy runs as a spectrum from the individual outwards to the entire city.

Extending the House into the Urban Area

In rural and road dominated suburban contexts each house is a single private destination in the landscape. Any journey by guests to an individual home is more than likely going to be a single purpose trip and not involve travelling to any of the neighbours. There is no public activity outside the front of the house other than passing vehicles. In everyday reality almost all activity and social interaction engaged in by the occupants of the household, whether old or young, has to be reached by means of a car, whether visitors coming to the house or the inhabitants travelling to some other venue (residential or other). From an urban design point of view, in rural and outer-suburban contexts the house can be construed as an isolated item, linked to the rest of humanity by a road – a single dimensional object. For the designer, thinking beyond the garden wall is essentially irrelevant. All design is contained to the property boundary and effectively operates independently of whatever lies outside that boundary.

From the perspective of the rearing of children, the house in garden concept works well in the rural context (including the isolated hamlet and remote village). As they grow, children begin to roam beyond the garden wall and are able to explore an incrementally wider and wider area of countryside behind and around the home. But in the modern car-dominated suburban context, where gardens and roads fill the space of the landscape, the accessible space for the growing child is strictly forever limited to the garden. The only legal escape from the private domain is straight onto the road – a highly dangerous, hostile environment. Unless cars have been suitably tamed through very careful masterplanning, then from the perspective of the growing child the influence of the car has made the residential landscape essentially dysfunctional. The consequence is a phenomena of the modern middle class world – children end up having to be chaperoned by parents for each and every social engagement; there is no or little scope for the child to do what is natural, to begin to investigate gradually wider and further afield from the home, exploring the social and physical world around on their own or just with their peers.

The only saving grace from this predicament for the maturing child is that in semi-rural and outer suburban contexts gardens themselves are frequently large, enabling significant roaming within a property boundary. But when housing estates are built entirely of roads and no streets with small homes and poky gardens, the only resort for the child is to become sedentary and focus their explorations onto the virtual universe of internet and video games. In this way, the car and modern housing estates will and are having an immense (quite probably, deleterious) impact on the health and maturation of the younger generations.

It has already been noted in the last chapter that there exists a threshold of increasing density where the highway in front of the house begins (or should begin) to be public space, as well as an access route. It is at this point that urban design as a discipline starts to have relevance. And it is at this point where it begins to become important to extrapolate the notion of the home, being structured according to degrees of privacy, into the surrounding landscape.

In the successfully functioning urban area the notion of degrees of publicity (looking outwards) should continue outside the individual home to embrace larger and larger communities of people: the family, the group of immediate neighbours, the neighbourhood, the community and so on. As with houses themselves, the physical structure of local urban areas does not need to manifest as concentric circles; but it should seek to achieve conceptually the same effect; each home should figuratively lie in its own least accessible location. Unfortunately, it is this extension of the concepts of privacy, security and containment, as homes are aggregated together to create the neighbourhood and beyond, where the design of housing has proven and continues to prove to be most flawed and problematic (at least in the UK context). In the next chapter consideration will be given as to how this appreciation of the role of the home should influence wider urban form; in the remainder of this chapter, the implications of this line of thinking will be considered as they relate to three particular aspects of housing design: inside the dwelling, low-density suburban masterplanning and high density design.

The Internal Design of Dwellings

The structure of the modern house is the result of centuries, if not millennia, of trial and error. It is not by accident that it does not matter where you are in the world, what local cultural influences there may be, religious or otherwise, the structure of modern detached and semi-detached housing is effectively universal. Whether an architect is trying to design a 10-bed chateau for a multi-millionaire or an affordable cottage with the tightest of budget and minimal space, the underlying template for the internal structure remains the same; it is driven fundamentally by this notion of degrees of privacy. There are, however, some additional influences which are worth noting and which have contributed to this iterative process to lead to the ideal home.

Most of these other influences are quite animalistic. Sleeping upstairs, for example, might give us a better sense of security, in part because being raised above ground provides a greater visual horizon and reduces anxiety on hearing predators, something which our evolutionary tree-inhabiting ancestors would have sought. Safety, of course, is absolutely paramount when it comes to housing; but safety and security cannot be used to explain why and how the modern home has become structured the way it has. Other factors such as heating and lighting may have historically had a greater influence on the evolution of the internal structure of dwellings: for example, requiring the hearth to be located centrally. However, with access to plumbing, gas and electricity and with the use of modern insulating materials and double glazing, these influences have fallen by the way-side.

It is this concept of degrees of privacy, which enduringly has the greatest influence on layout of homes. The degree of formality between family and wider society will, for example, inform the desire for reception rooms, which are separate from a family's normal living, eating and cooking space; but such differences represent added cultural expectations onto the underlying template of degrees of privacy. Anyone grasping this notion and thinking through how it might operate in their own cultural context should be readily able to design an appropriate structure of a home peculiar to and pertinent for their local cultural environment.

It would seem simple enough. But it can still so easily be done wrong. Here is an anecdotal observation. Housebuilders in the UK have a propensity to stick kitchens and dining areas at the front of homes and put living rooms/reception rooms opening up onto the garden at the back. Yet when people design their own extensions or adapt their homes for themselves, time and again they put the living room/reception at the front of the house and put the kitchen/dining area at the back. There are two motivations for such design choices by homeowners.

A living room at the front of the house can more easily be treated as a reception room. When a more formal or less familiar guest visits, they can be ushered into such room without passing through more private parts of the house. Preparations for tea and cakes can then take place back stage, just how the Victorians liked to do it (dirty crockery, waiting to be washed up in the kitchen, can remain unseen). Working in concert with this, placing the kitchen/diner at the back, opening directly onto the garden, enables the parents of the household to observe better the activities of children in the garden when the parents are doing the cooking or washing up. Such structure works better in terms of providing safe, observable containment for children, in which to play and explore. Garden as play space is more important on a day-to-day basis than garden as reception space. It is consequently quite bemusing that housebuilders, who are otherwise so adept at appreciating the needs of their customers, get this so frequently wrong.

Here is another anecdotal example of this principle going awry. A developer of a highly prestigious, high rise London development failed to sell a single flat to the

Far East market in a particular high-rise (35+ storey) tower. The simple reason: in every flat, the first thing you saw on opening the front door was into the bathroom. There is a direct line of sight from front door into the interior of the bathroom, one of the most private parts of any dwelling. One need be no expert of *feng shui* to appreciate that this would not be desirable to discerning purchasers.

In the context of the detached and semi-detached house, the internal structure can be dictated purely in terms of providing maximum functionality for the occupants, driven by the prevailing factors of privacy and containment for children. As dwellings become packed more closely together, then (as noted in Chapter 7) the surrounding environment begins to have a stronger and stronger influence on how the internal parts of a home are laid out; but the internal principles remain; they simply need to be married thoughtfully with the outside world of surrounding other homes, communal and public spaces. For a much more detailed exposition on how the internal elements of housing need to change as density increases, then consider David Levitt's book *The Housing Design Handbook: A Guide to Good Practice* (Levitt, 2009).

Suburban Masterplanning

Drawing on the ideas earlier and within this chapter, it is an informative exercise to dissect a very modern, very typical urban experiment – the urban tree. Urban trees are those large housing estates, produced invariably by private sector housebuilders, where there is a singular entrance road to a large housing estate, which is structured like a tree – with branching side roads leading to ever smaller branching side roads, eventually terminating in cul-de-sacs.

When Christopher Alexander wrote his famous thought piece *A City is not a Tree* in the 1960s (Alexander, 1966), he had not imagined how far the tree paradigm might go. The cities, which he complained had been designed as tree-like fractal structures, were nothing like the simplified trees, which have been created in the UK and US suburban landscapes over the last few decades. Alexander's mathematical discussion on the inability of designers to comprehend and design in the real complexity of urban landscapes never envisaged that some decades later many suburban areas might actually be designed literally as tree-like structures. Thinking in terms of the difference between roads and streets and the true essence of housing (just outlined) enables a more thorough, and practical, appreciation of the faults of these recent suburban structures.

It should first be noted that the drive behind housebuilders producing suburban areas according to a tree-like template is not to create great functional places for people to live. Far from it. In a typical new suburban development (even relatively large ones), the maximum value that can be achieved for any new house is dictated by the background values of the existing housing stock within the local urban area. If a housebuilder is seeking to build 3-bedroom houses, then his agent will peg the value of these new 3-bedroom houses with any alternatives that someone looking for a house might be able to acquire within the vicinity, adding a small uplift for the new build. So, for any suburban or semi-rural developer the way to extract maximum profit from a new development is two-fold: by minimising costs and by maximising density. Value plays very little part in the thinking.

There are two principal ways, in which the suburban housebuilder can minimise costs. One route is to be efficient in construction of the houses themselves, to produce identikit boxes, all designed and constructed in exactly the same way, where parts can be ordered in bulk. The other way is to minimise infrastructure costs. This can be achieved by minimising the total length of roads and sewers, which need to be constructed to serve the new boxes. It is not by accident that nature makes use of tree-like fractals in many situations – trees, lungs, capillary systems, etc. – where maximum surface area is required for minimum length of 'pipework'. Likewise, the tree-like suburban estate maximises the number of dwellings with access directly onto a road with minimum total length of tarmac and underground pipes and wires.

In order to extract more profit out of the land, given the cap on individual box values, the developer needs to squeeze as many houses on the land as possible. To keep the values of the individual houses to the highest possible level, each box must be detached (in England we seemingly put a significant extra value on detached verses semi-detached or terraced housing). This also makes it easier to build identikit boxes. These houses are then built at densities, which should normally be semi-detached, even terraced. So none of the benefits of these latter two housing forms are achieved in terms of preserving and maintaining usable external amenity space. The outcome is an estate of boxes built higgle-dy-piggle-dy on a tree-like road structure, where each house is constructed to a standard template and where, whilst in theory each plot may be of a seemingly reasonable size, the usefulness of much external space is lost by being tucked down the sides of houses, used as redundant front gardens, or in odd-shaped, contorted and overlooked rear gardens. What the homes in these estates desperately need is some extra amenity space – just the sort of thing a good residential street could provide. But …

The tree-like suburban estate creates a number of cul-de-sacs, which do indeed create good residential conditions in the sense that these houses are in very (very) secluded locations, away from the hub-bub of the town or city, and do not provide through routes for everyone else. The cul-de-sacs themselves can (if designed well, though frequently not) act as public space for those living around them and provide additional amenity and recreation space for a few houses. In theory, children can spill out of the houses and play in these cul-de-sacs without fear of being run over. Looking at the rest of the highway network, with the whole estate formed around a tree-structure, by definition all other roads are indeed roads, with ever-increasing levels of traffic funneled along them to reach the single entry/exit for the whole estate. With the exception of those few cul-de-sacs, no other part of the highway network through the estate can act like streets or contribute to public space for the residents.

For people, especially families, choosing to live on these modern estates of boxes with poky gardens, they have a difficult decision to take. On the one hand they can choose to live on a cul-de-sac, where they might gain some usable amenity, but be in such an inaccessible location that every journey has to be by car; as the crow flies they might be relatively close to shops, amenities and anything else; but along the road network, they may need to travel miles to get to the nearest corner shop (or equivalent). On the other hand, they can choose to live closer to the entrance of the estate with better accessibility to the outside world, but suffer the consequence that everyone else further into the estate must drive past their front door; the road, on which they live, will consequentially deter the creation of any local community and the road, being a road, will not confer any desperately needed additional amenity space.

Earlier in this chapter it was suggested that the ideal residential neighbourhood is designed in such a way that every dwelling is located at the centre of its own private universe, surrounded by a community of other homes, gradually giving way to a more public outside world. Yet at the same time each home needs to have reasonable access to facilities and amenities. The tree-like estates fail these objectives spectacularly.

Such suburban estates appear to be a peculiar Anglo-Saxon phenomenon, which occur in the UK, certain parts of the US and in some international enclaves with strong UK or US influence. Where land pressure is less intense, then at least housing plots themselves tend to be large enough to confer decent private external amenity and good sized houses; it is therefore less critical that there are next-to-no streets in such residential areas; but they do mean that residents have to drive miles, sometimes many miles, to reach any facilities or amenities. As (over the long-term) fuel prices continue to rise, the ends of the road tendrils into these branching housing estates will become increasingly isolated and costly to live in.

There are much better solutions to the creation of residential areas. The overall form is a different type of fractal. These urban forms were once much more common in the UK (some long-time past, long before the car), when the drive behind residential development genuinely sought to create places that people may value as a place to live and where accessibility by foot to a market place was balanced in equal measure to the desire for privacy. These urban forms continue to be more common in Europe, where there tends to be much greater public sector control of the early masterplanning, where urban planning is driven by a genuine desire to create good places to live rather than from the short-term financial imperatives of private sector housebuilders.

Figure 9.1 shows some alternative fractal structures, which generate streets and homes which are suitably secluded from busier parts of town but not inaccessible or isolated. In some cases these patterns rely on cul-de-sacs and in some cases the smallest streets are mews. These structures do not force people wishing to live in the area to make Faustian choices of one disadvantage or another. They create urban environments, which are relatively equitable – each home has relatively similar levels of privacy and accessibility and fronts onto similar degrees of public space – most highway is closer to being a street than a road. The permeability of the urban grain means that traffic is spread out throughout the urban landscape, allowing each section of highway to be both road and street.

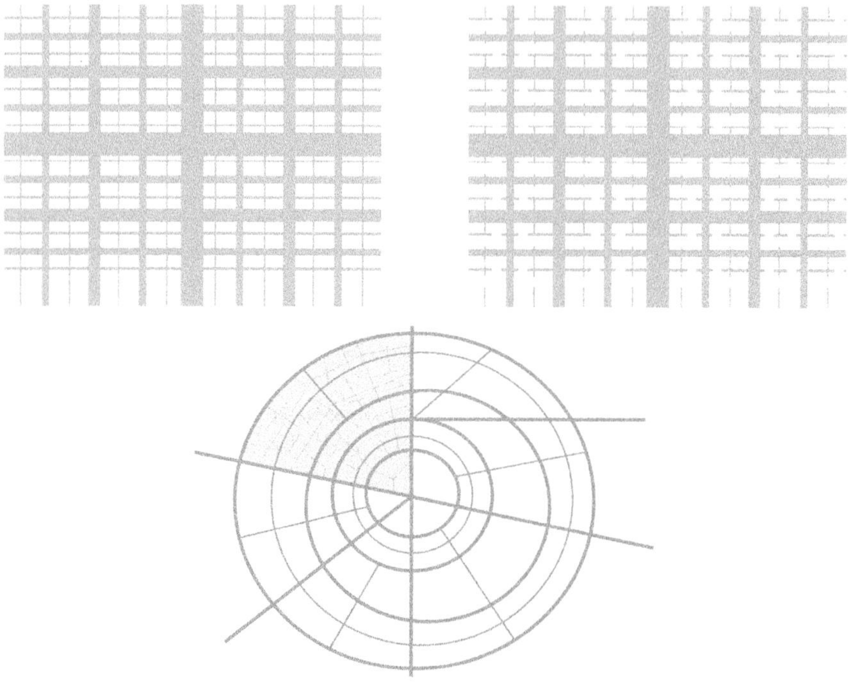

Figure 9.1 Town structures that can provide streets

In all these solutions there are always some difficult choices, where some carriageways are always going to be busier and more road-like than others. But in the European preferred solutions it is easier to locate appropriate buildings onto these stretches of road – for example, blocks of flats can be located on the busier routes, with smaller terraces of homes on the quieter sections of highway. The natural evolved solution can be seen en route into many older market towns and cities, where large detached homes with larger gardens are sited on the main arterial roads into town and smaller homes are located on the cross-interstices – see Figure 9.2.

The relationship between what has been set out here and the latest thinking and best practice in urban design, such as By Design (CABE, 2000) will be discussed in Chapter 13.

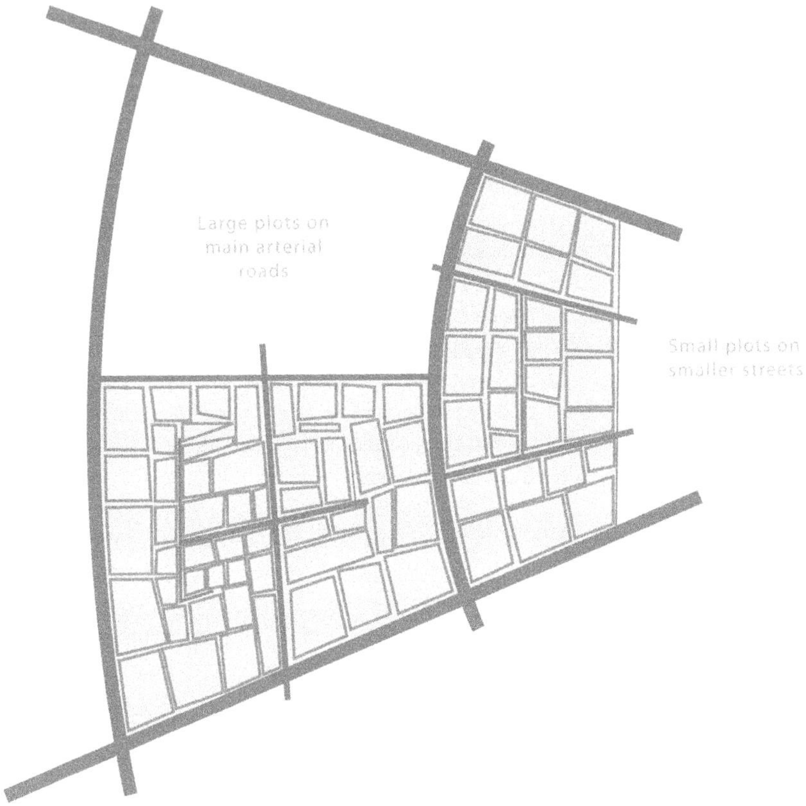

Figure 9.2 Plot variation for streets and roads

High Density Solutions

In London in the last decade a similar problem arose to that just discussed with respect to tree-like estates, but in a considerably different context.

Before the turn of the Millennium, high density housing and especially anything high rise (that is more than four storeys) had become a complete anathema in the UK. The failings of the 1960s and 1970s estates had created a severe backlash against anything perceived as being high density (whether or not it was actually high density). Housing pressure in London[3] and the drive to re-use previous

3 When the Greater London Authority was established, they were tasked to produce a London Plan (Greater London Authority, 2004), which amongst other things needed to deal with the huge housing pressure facing London. This led to a relaxation of the planning rules with regard to density and the first appearance of the London Density Matrix, which links density with public transport accessibility levels.

industrial land[4] caused a sea-change and a fast and furious revolution in how we approach urban housing. But after over 30 years of building only suburban and low-density, we collectively had very little experience either how to design high-density housing (developer's perspective) or how to plan and control new high-density housing (from a local and national government perspective).

In the centre of London and other major city centres, a gold-rush ensued as planning controls were relaxed and landowners and developers sought to cram more and more flats onto plots of land. For many small to medium size developments, the same value and cost dynamics applied as for the suburban masterplans. The value of individual units is capped by the background apartment values in the local urban area, determined in large part by what type of area it is in socio-economic terms. So, for the average development the way to maximise profit is to maximise number of units on a plot of land and minimise cost of delivery. The design process, rather than seeking to create great places to live, was wholeheartedly focussed on solving these quantity and efficiency challenges.

With a bit of trial and error, leading urban developers soon discovered how to maximise building efficiencies of blocks of flats. This involved minimising the number of cores (i.e. lifts and staircases) and designing each floor to have one long internal corridor with single-aspect flats on either side. These super-efficient buildings can now be seen in many places around London. It quickly became apparent, however, that they were problematic in terms of both their impact on the wider urban landscape and internally.

Considering first the impact on the wider landscape, if you take a row of terraced houses or a semi-detached houses and replace them with a single block of flats with only one pedestrian entrance and one car parking ramp, then you are immediately beginning to change the character of the street. Where once there were frequent comings and goings from numerous doors directly onto the street, enabling interaction and activity across the street, there is now a single entry point within an otherwise blank façade. The relationship that local residents have with the public realm and the types of interaction that may take place within it has, all of a sudden, undergone a marked change. Most notably, drawing on the ideas earlier in this book, the activity will become much more directional along the length of the carriageway rather than across its width or between immediate neighbours. What was once a street begins its transformation into a road … and all that flows from that.

Recognising that the financial dictats of the industry were driving building and urban design in a negative direction, the design industry has fought back. The London Mayor's London Housing Design Guide embedded thinking to seek to counteract these changes. The guidance is beginning to show fruit through the New London Vernacular (Hart and Birkbeck, 2013), where many schemes designed since 2010 have sought to re-introduce street-based thinking into new developments. Where once a block of flats had only a single entry point, in good

4 As evidenced through the work of the Urban Task Force (UTF, 1999).

designs the ground floor units are frequently designed as maisonettes all provided with access directly onto the street. The buildings themselves may have grown and the density significantly increased, but the urban design solution keeps the street and maintains the potential of transverse activity within the public realm.

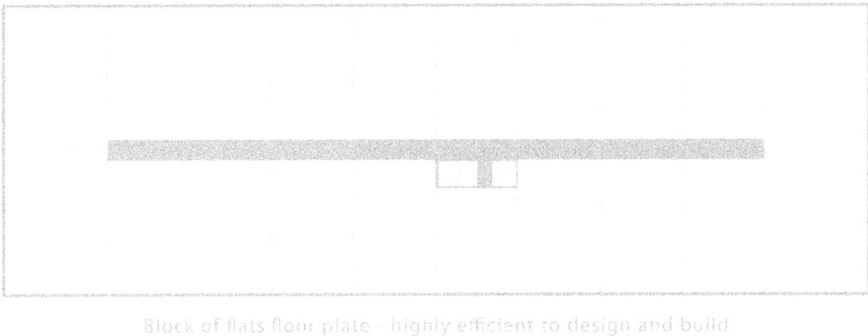

Figure 9.3 Floor plan for a highly efficient residential building

Within the highly efficient new blocks of flats there have been equally problematic issues (see Figure 9.3). In particular, the long internal corridors represented an internalization of the deck-access/streets-in-the-sky concept, so often characteristic of the 1960s housing estates (see Box 9.2). These long internal corridors may have been good for corridor cricket (a natural solution from the child's perspective, where the flats had no real play space whatsoever and balconies which were sufficient for no more than a cigarette break). But where families and young professionals have been pepper-potted[5] together on these long corridors, then the resultant (and inevitable) social conflict and anti-social behaviour was highly predictable and really should have been avoided.

Box 9.2: Deck Access – acceptable or not?

An idea, which became manifest in physical form during the 1960s, was the proposition that to deal with the motorcar there should be physical separation between pedestrians and traffic. This influenced the design of many city centre areas in the post-war period and the new emergent public housing estates. In the context of the housing estates, this became expressed in the form of, what have become known as, streets in the sky. The concrete towers of these new estates were connected by continuous pedestrian decks, which could run for a kilometer or more, connecting a series of buildings.

5 Pepper-potting is a term used within the housing industry where flats designed to accommodate different tenures of housing or people with very different lifestyles are mixed together within one block of flats.

Alice Coleman (1985) documented the problems that these design solutions produced in social terms. Quite simply these high level corridors fail dismally with respect to any notions of privacy, community or defensible space. They were normally freely accessible to the general public and anyone else on the estate, yet they were not overlooked in any way. Anyone walking along these long decks could do so with complete anonymity. As a consequence a single woman on her own in a flat would have no idea who might be knocking on her front door; and when she opened it, the visitor could force his/her way in, do whatever they want, with impunity, knowing that no one, nowhere was or could be observing their actions. This contrasts with a normal street, where anyone walking along it will (or at least in theory could be) observed numerous times and anyone knocking on a front door might be observed by other pedestrians walking along the street at any moment.

The streets in the sky were neither public nor private domains and this uncertainty created severe social problems and gave rise to rampant crime.

The highly efficient new blocks of flats built in the period 2000 to 2010 were replicating the same problems, but internalized. The long external decks were now internal corridors in buildings of several hundred flats. Again, anyone living in any flat would have no idea who might be knocking on their front door; and when opened, that visitor could do what he pleased knowing that no one could be observing them. As a result Secured by Design guidance, consistent with the London Housing Design Guide, now suggests that in blocks of flats there should be no more than 10 dwellings behind the last layer of security. This should not be read as a definitive rule (10 is a fairly arbitrary number in this case), but as a guideline level.

The key message to take from this is that the internal design of buildings should be treated as an extension of urban design. The ideas proposed within this chapter with regard to layers of security/privacy apply at all scales of the urban landscape from deep within the private dwelling extending outwards to encompass the whole city.

As noted above, these highly efficient buildings necessitated single aspect flats. There is nothing necessarily wrong with a single aspect flat, when designed well with a good aspect; but when designed badly on a north facing façade with no view to speak of, then such flats are far from ideal. Equally problematic are poorly designed single aspect flats on south facing facades overlooking noisy roads, which can quickly overheat with disastrous health and psychological effects.

The *Interim London Housing Design Guide* (Greater London Authority, 2010) has sought to address these problems by limiting internal corridor lengths, promoting security on each corridor to create small communities with access to each floor of a building and making sure that single aspect flats are only provided in agreeable conditions conducive to creating a comfortable internal environment.

The appearance in the last two years of a possible new vernacular for housing across London suggests that, after a decade of trial and error, leading housing designers across the capital are beginning to learn (from their own experience and from each other) how to design high-density housing that works at both an individual level and for wider communities. There are still many challenges ahead, especially in terms of achieving very energy efficient homes and the lifestyle implications. And there are still mistakes made on numerous occasions. Hopefully the ideas and principles rehearsed in this and previous chapters can help provide a better urban design framework to guide professional designers to produce holistic solutions that work at all scales, from en suite bathroom to high street.

Summary

The essence of all housing is privacy. Understanding this helps to explain the common template for the configuration of homes across the developed world. Such structural form has been reached over generations of trial and error to produce something that generally works for people. The concept of privacy should, however, extend beyond the interior of the home into the wider urban structure; this has major implications for how we should seek to structure residential areas. Unfortunately, in the UK, whilst we are quite good at designing individual homes, for many years we have failed dismally appreciate the importance of extending the concept of privacy outwards beyond the home into the wider urban area.

Chapter 10
A Theory of the Town

The dictionary definition of design as a verb runs along the lines "prepare the preliminary sketch or the plans for …, especially to plan the form and structure" or "to plan and fashion artistically and skilfully". Whilst these definitions might be technically correct, they seem to fall somewhat short of the experience and reality of design in the context of masterplanning new urban developments in the modern world. The latter may indeed lead to the creation of sketches and plans for the new physical form of an area of the city, small or large; but the entirety of the design process for larger projects is more akin to a collaborative problem solving process than simply one of drawing up plans. In times past the architect or masterbuilder had to solve the problem of gravity; but apart from that he was largely left to his own devices to draw up plans of what he wanted to create. Designing in the twenty-first century requires the designer (or more likely design team) instead to resolve a multitude of different demands, whether generated by the physical environment (contain some contaminated soil or avoid a major electrical line) or legal constraints (cannot build above a certain height) or social and cultural expectations (the local community wants a new community centre).

The design team in the early twenty-first century has various challenges to overcome, not least of which is that (as mentioned in Chapter 1) the discipline of urban design lacks any cohesive theory, which might guide a designer along a path to an optimal solution (Carmona, 2003). In times long past, several centuries ago, there were great architects who intuitively saw how the city worked, how people related to space, and they created great streets and piazzas accordingly. These are places, which have lasted the test of time. Over the course of the twentieth century, the world of architecture has evolved through a series of ideologies – the garden city movement, the mechanistic city, the modernist city, the brutalist movement, and so on; each of these represented guiding theories on how to design a city to solve the social problems of the day; and each of these ideologies has been found wanting in various ways.

With the codification of the environmental movement into European law, in particular through legislation around environmental assessment, the design process in relation to the built environment has embraced a new ideology: it has become much more scientific. The task of the architect, nowadays inevitably working under the banner of sustainability, is to understand context. He or she needs (or at least is meant) to research every nuance and facet of the social, economic and physical landscape with which he or she is working, and then design in response to that context. We no longer have any guiding ideologies as to what to achieve; but we do have an ideology concerning how we should go about the process of design. Perhaps this is simply a counter-response to the previous sequence of failed ideological movements.

Once the architect has been provided with all the parts of the jigsaw puzzle by the numerous other technical consultants, from air quality and acoustic specialists to cultural advisors to civil engineers marking out where all the existing utilities lie, then he or she is expected to fashion skilfully (and in collaboration with those technical specialists) a plan for a regenerated area of the town. Information overload! All this knowledge about an area disguises the real truth: a serious lack of understanding.

The inconvenient truth in the world of urban design is that there is no comprehensive theory about how the town should work or how people (individually or collectively) relate to space, or how economic drivers inform the form of towns and cities. That is not to say that there is no theory at all; there are many sub-theories and theories about bits and pieces of the urban puzzle or which are very specific to a particular city or region; there simply is not a singular theory which brings all the different strands together and that actually seems to work. Instead, students are taught how to go about the practice of urban design and all the things that they should take on board and consider during that process; they are then told to go forth to create great *places*, without anyone ever genuinely seeking to explain exactly what it is that makes a *place*, apart from showing them some examples of spaces that are deemed to be great places.[1]

Hopefully the pages in this book heretofore have provided some better insights as to how our towns and cities really do function and how form follows function. Through this chapter an attempt will be made to bring the various strands of thinking together into a coherent picture.

What is the Problem to Solve?

Drawing on the ideas presented through the earlier chapters, it is now postulated that, underlying all the complexity of the urban environment, when considered from the perspective of *form following function* there can be construed to be *three major drivers of urban form*. These drivers have (in a sense) an order of priority and they shall be tackled in that manner; they are each summarised briefly first and then considered in more detail.

First and foremost is the need for the city to operate as an economic entity to feed the inhabitant population. Goods (especially food) must be brought into the city and then distributed to the population of the town. Adam Smith, some 300 years ago, observed that the human solution, which has been found most effective, to facilitate distribution of goods from those who grow, make or source them

1 The one exception to this statement about lack of theory is ideas around space syntax (see Box 10.2). More discussion is provided in Chapter 12 on the correlation between the ideas presented in this book and space syntax thinking.

to a human population of customers, who need to consume or use those goods, is through the free market. It was then established in Chapter 1 that the most energy efficient way that this can be done is by locating a physical market in the gravitational centre of a human habitation, creating the market place, and thereby establishing a spatial relationship between the different land uses.

This suggests that the most important movement patterns for any town are the arterial routes into and out of town. Without carriageways to enable such movement a town cannot function properly on a day-to-day basis. These arterial routes are necessarily roads, connecting the centre of a town to distant urban centres and to the landscape around the town. As explored in Chapters 2 and 3, if these roads become major motorways (or railroads), then they can end up quartering the town into slices of a cake, preventing the disparate parts of the town from communicating between each other.

This brings us to the second major driver of town form – the need for other movement patterns and lines of communication around the town. In the theoretical extreme case that the arterial roads into and out of town become impenetrable because of high traffic volumes and speed of traffic,[2] then all communication between any part of such city could only take place through its centre. The urban fabric of the town would have been turned into a cake, which has been sliced, where no communication can take place between the different slices of the cake unless this takes place via the centre of the town. This may work fine for a small town, but as any urban centre grows towards a city, then the centre of town would become chocked up with internal town traffic trying to link with other parts of the town. At the very least the town would not be functioning efficiently.

This brings into play the second direction of accessibility that is required for a town or city to function successfully. Economically a town's success is also dependent on circumferential movement between the *slices of cake*, around the town. If this were not to happen, then the town may as well be six (or so) smaller towns, each slice of cake operating independently. The town would lose any economic advantage in scale terms (through, for example, agglomerative economics) arising from the whole population living in close proximity. Enabling cross-movement around the town facilitates both economic and social interaction. In economic terms, businesses from one part of the town can do direct business with each other without having to do so *through* the town centre (*through* could here mean both physical movement through the town centre or via a third party operating from the centre of town). In social terms, communities around the town can better interact with each other, enabling greater cross-communication of ideas and sharing of amenities and facilities.

As set out at the beginning of Chapter 9, these two main drivers of accessibility (arterial and circumferential) do not, however, explain the full story. At the much

2 Laughable as it may seem, there are many large cities around the world where major urban motorways have cut off large urban areas and isolated their communities from the rest of the city.

more local level, there is a third equally important driver, which is actually to reduce accessibility, to create privacy and seclusion for residents within their homes and very localised communities. At the smallest scale of the townscape, the need for accessibility gives way to a need for privacy.

To facilitate the following discussion, these drivers of urban form have been labelled as follows:

- Driver 1 – Purposeful Movement – *enabling radial accessibility*
- Driver 2 – Social and Economic Interaction – *enabling circumferential accessibility (in any event orthogonal to Driver 1)*
- Driver 3 – Privacy – *limiting accessibility*

As will hopefully be immediately obvious, these three drivers operate antagonistically to each other. Improving access into and out of the city must be balanced against traffic needing to travel in a perpendicular direction around the town. In larger towns, both of these drivers for improved accessibility can begin to give rise to roads, which could become detrimental to the creation of privacy at the very local level.

Each of these drivers of urban form represents the social and economic activity of the population of the city becoming expressed and frozen onto the physical landscape of the city. Where enough people choose to move along a particular route, for whatever reason, then over time a footpath becomes a lane becomes a carriageway, all the way up to a ten-lane motorway. Human actions take physical form in the highway network, at the macroscopic scale, and in terms of building forms and configurations at the micro-scale. These three primary drivers continually act against each other, creating tensions in urban development and regeneration; where one or other comes to dominate (by accident or by intervention from planners and politicians), then this becomes displayed in quite predictable ways across the townscape: for example, the arterial urban motorway facilitating trade between the city centre and the outside world, yet simultaneously creating an impenetrable barrier through the town, dividing once cohesive communities and well-functioning business precincts. These drivers are very real and ever-present in every human town and city.

The challenge for any urban planner or designer is to understand how these drivers and the tensions they create are actually expressed or should better play out across the area of town, which he or she is seeking to change, improve, regenerate or fill in.

Driver 1 – Purposeful Movement

As already stated, this is radial movement in and out of the centre of a town. As rehearsed in Chapter 1, a town, any town, is first and foremost an economic entity; to exist and survive from day-to-day a town must at the very least receive goods (food and other materials) into the town and these goods must be distributed to the population of the town. It is also critical for the town to be able to transport out its waste materials. Any larger town should also be acting as a trading centre, exporting manufactured or locally resourced goods and acting as a source of materials to its wider rural and industrial hinterland.

Focussing on the minimum fundamental requirement of food, this must *daily* be brought into the town and *daily* be available for purchase by the townsfolk. Transport of food and associated necessary materials must flow as smoothly as possible into the town and then to all parts of the town, driven as we know by the free market and Adam Smith's each and every person acting out their own and their family needs.

This flow of goods creates primary access corridors, running radially into the centre of a town. Where a town has multiple centres, such routes must also run into and out of (or through) the centre of each of these subsidiary urban centres. The importance and daily urgency for food to be unhindered is the reason why, at the macroscopic scale, this driver is paramount.

In the context of a small market town, the need to provide fast and efficient radial transport infrastructure does not unduly inhibit other parts of the town. The flows of human and goods traffic are not so great that such primary access corridors become barriers. But, as considered in Chapter 2, when the size of the town is scaled up to a city, these primary access routes transform into major arterial motorways or railway lines, which cut swathes through the urban landscape. They may facilitate directional and purposeful movement directly into and out of the town, but they become impenetrable barriers to any other direction of movement, and thence interaction, around the town. As a consequence businesses and communities become separated and cut off, isolated from the outside world of the rest of the city and beyond.

It was the creation of such primary corridors, through the building of new motorways, and the impact this had on erstwhile cohesive communities, which so riled Jane Jacobs and drove her to write her seminal book *The Death and Life of Great American Cities* (Jacobs, 1961). A city, which has allowed the primary access corridors to dominate, becomes internally dysfunctional. It may be very successful in the import and trading of goods with the outside world, acting as a major marketplace; but it becomes very poor at manufacturing, simply because different parts of the city cannot communicate with each other. A business operating in one quadrant is effectively prevented from cooperating with businesses in other quadrants because of the time taken to travel from one premises to another: in the extreme only via the city centre. This is not to say that a city, whose primary access corridors have become major barriers to movement around the city, cannot function as a manufacturing centre; rather it will simply be much less efficient and effective in that capacity and less able to compete with other cities.

Driver 2 – Social and Economic Interaction

For a town or city to operate successfully in any way above and beyond that of being purely a trading centre, then it must be structured to enable fast and efficient interaction of people and businesses throughout the whole townscape. All parts of the city need to be reasonably accessible to all other parts of the city.

In Chapter 3 it was elicited that roads and streets lie at different ends of a spectrum, where that spectrum is defined according to the degree to which a particular carriageway manifests and facilitates longitudinal or latitudinal movement. There is an inherent tension between these perpendicular purposes of the carriageway: on the one hand to connect distant destinations (town centre and outside world) and on the other to connect the local townscape. The larger superstructure of the city is simply a bigger expression of these orthogonal tensions operating along each and every road/street, especially those major arterial routes connecting town centre with the outside world. Where arterial roads meet circumferential routes, there is clearly a juxtaposition of objectives: to serve the city centre or to allow the rest of the town to inter-communicate.

There are various ways to deal with this, which may depend upon the urban density of the city – see Box 10.1. In very large cities, where there are the resources to make very significant capital investments, then the arterial routes can be jacked up, undercut, buried and regularly bridged, to prevent them acting as barriers to more local movement and interaction. This does, however, isolate such macroscopic movement patterns from the local suburban parts of the town. Alternatively, these arterial roads need to be slowed down at the junctions with circumferential routes to create streets, which can intersect and interact with the more local. This can then give rise to the creation of suburban centres that properly connect into the local parts of the city. Such solutions, however, clearly create competing tensions across the urban landscape. Streets along these arterial routes will impede the rates and flows of traffic into the main city centre; the more street-like these feeder routes become the better they will serve the suburban centres, but the more they will impede the through traffic. This effectively sets off the suburban centres and main city centre as competing entities, where a reasonable balance has to be achieved between routes serving local centres and accessibility to the main city centre. This tension exists across a variety of scales, wherever there exists a hierarchy of town centres. At the end of Chapter 3 consideration was given on the complexity of achieving such a balance across a 2-dimensional grid compared, for instance, to along a single dimensional railway line.

Box 10.1: Low-Density and High-Density Resolution of Competing Drivers

Low-density and high-density cities deal with the tensions between the two main urban drivers in very different ways, leading to different urban super-structures.

In low-density contexts, the urban landscape becomes dominated by roads. Both the arterial and circumferential routes through the town operate as roads; crossings of these routes will more naturally be designed as cross-roads. Manifestly, cross-roads are not conducive to the creation of any suburban centres. These two sets of roads will act together to slice up the urban landscape into isolated, irregularly shaped blocks. The only way that anything sitting within these cut off urban areas can interact with the outside world is by means of further sub-roads, feeding onto and off the major highways. Possibly immediately and certainly over time these cut off urban areas are likely to become specifically characterised – one low value housing, another high value housing, another a residential area dominated by a particular religious group, another a retail park, another a business park, another becoming beset by crime, etc. There is no real incentive for the creation of any significant sub-centres of the city; there will be the major city centre of skyscrapers and everywhere else will be made up of low-rise private destinations (i.e. Malls, business parks, housing estates, etc.). And everyone must travel anywhere by car.

In the high density contexts, the solution is for the arterial and circumferential roads both to slow down at crossings. This can create the opportunity for the appearance of streets at these junctions, which in turn can lead to the formation of sub-centres at these locations, drawing activity to them until significant suburban town centres are formed. The main arterial routes end up operating like a railway line – fast, slow, fast, slow. The resultant sub-town centres manifestly operate in tension with the main city centre, but in turn lead to the formation of a more balanced urban map (an ecosystem and hierarchy of town centres, all competing and cooperating across the urban landscape). It is a much more interesting urban map in terms of urban diversity – a genuine urban ecosystem. With the formation of local town centres, these connect better into the local urban map of residential areas. And the connected residential areas, being connected into truly public local town centres, are naturally going to be more mixed in nature – socially, economically, etc.

These two urban forms, the low-density and high-density, are arguably mutually incompatible. They cannot successfully function in the same place, within the same part of a city, the same distance from a city centre. To be successful, cities really need to choose to be low-density (below 50 dwellings per hectare) or high-density (above 80 dwellings per hectare), and not mix and match the densities.

Standing back and looking at the overall movement structure of the city, it is apparent that the combination of arterial routes and orthogonal internal city routes creates the classic spider-web pattern (see Figure 10.1). And it so happens that space syntax analysis (see Box 10.2 below) suggests that the spider's web is indeed the network pattern that gives rise to the maximum level of absolute accessibility[3] (or connectivity) to all parts of the urban landscape; the spider's web generates marginally more absolute accessibility than the classic grid pattern. Yet, pondering on this, it can also be seen that, away from its very centre, the spider's web creates an almost grid pattern; and the further from the centre travelled, the closer the spider's web approximates to a grid pattern (see Figure 10.2).

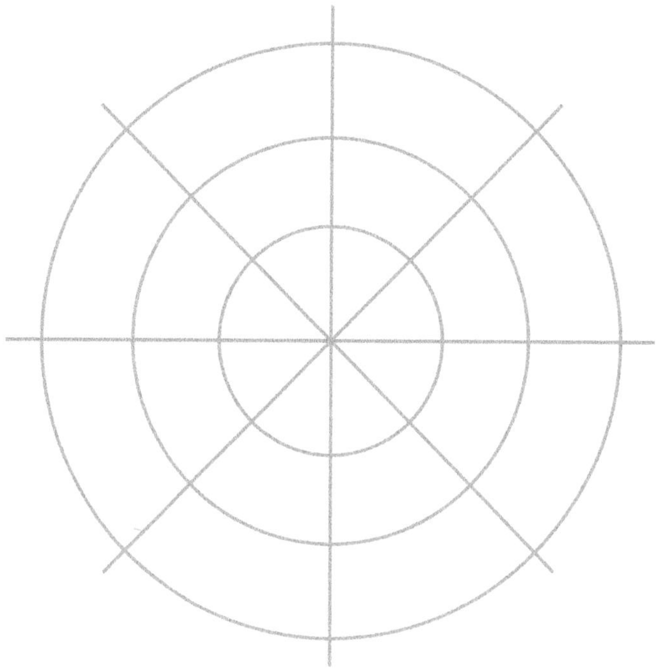

Figure 10.1 Classic urban pattern – the spider's web

3 Absolute accessibility is here taken to mean the pattern which best connects each location with every other location. Clearly the centre of town is itself the point of maximum accessibility to all other parts of the town, facilitated by the arterial routes all running to the centre. The spider-web pattern, however, means that all other parts of the town are also well connected to each other.

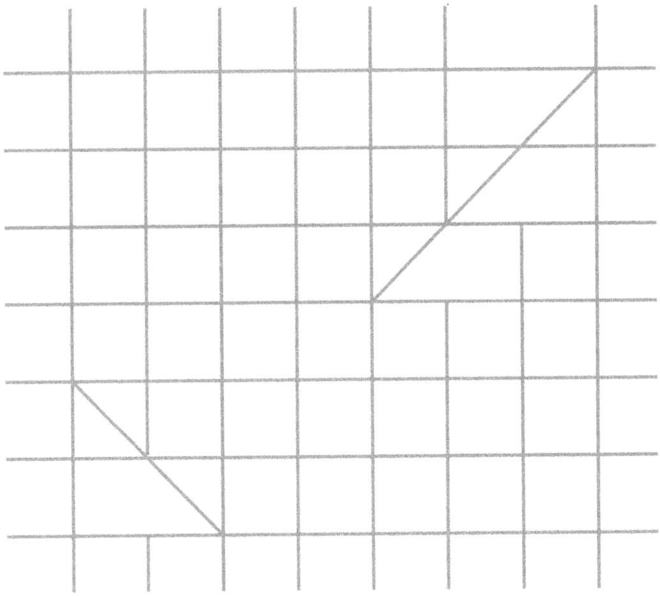

Figure 10.2 Classic urban pattern – the gridiron

Box 10.2: Space Syntax – Brief Explanation

Space syntax originated at the Bartlett School of Planning, University College London in the 1970s and 1980s. It includes a set of theories and techniques for the analysis of spatial configurations. It was first conceived by Bill Hillier and Julienne Hanson as a way to understand better how the layout of the built environment affects human behaviour and the route choices that people make, in particular when walking through a city environment. Early models of space syntax broke down the shape of city into a series of visible polygons based on sight lines and then predicted how people would naturally travel through those spaces, according to what they could see of the space around them. One of the core premises on which space syntax is based is that people will seek to travel from A to B with as few turnings as possible.

In due course this has allowed the construction of algorithms, which can be used to build accessibility maps of the urban landscape. Such maps can show how accessible each road/street segment is to all other road/street segments across the city. The algorithms can be set up to take travel distance into account or not. Without any pre-knowledge of actual movement patterns in any particular town or of the density of habitation across different parts of a town, space syntax maps are often more than 80 per cent accurate in showing what parts of a town are naturally more busy and active verses those that are quiet and less used. The level of accuracy arising from space syntax models gives strong support for the premise that form follows function, that the urban landscape arises as a consequence of the needs and patterns of movement of the inhabitant population.

Space syntax maps can be constructed at different scales, showing how well integrated a road/street segment is within the very local, the intermediate and the macroscopic urban landscape. A road segment would, for instance, likely be well integrated into the big picture of the city, but relatively isolated from the immediate locality; the reverse may be (or should be) the case with respect to a residential street.

Until recently space syntax methods have tended to be used and applied separately from other types of transport and traffic modelling, which typically have been built up through the gathering of empirical data on the actual travel patterns from and to defined destinations across the urban landscape. These approaches were criticised in Chapters 2 and 3; they reinforce a thinking in urban planning focussed on privatised destinations. This situation is beginning to change with space syntax gaining greater credibility in the transport modelling world and starting to be incorporated into those more conventional models.

Driver 3 – Privacy

Drivers 1 and 2 are most obviously expressed in the macroscopic and intermediate scales of the city structure. They can also be understood as being urgent drivers of the overall effective economic functioning of the city. The third driver operates at yet another (in essence) perpendicular direction to the first two principal drivers. Its orthogonality to the other two drivers is, however, more difficult to understand: for starters, how can there be a third perpendicular within 2-dimensional space?

As explored in Chapter 9, the quintessence of the home and everything associated with the home and socialising around the home, especially with regard to the rearing of children and formation of communities, is the enclosure of space. This acts as a localised counter-play to maximisation of accessibility. The layers-of-privacy concept clearly operates antagonistically to speed and ease of access.

Chapter 9 started with a discussion on the intrinsic structure of housing. For a home to operate effectively for its true purpose not only must it be structured to provide layers of privacy, but, walking out of the front door of the house or flat, the surrounding city should continue to operate in a (figurative) concentric ring manner, allowing the resident into areas of greater and greater publicity. Or, from the child's perspective, allowing him or her to explore areas of an ever-increasing public nature, until eventually as a young adult they are allowed free, unaccompanied access (by their parents or guardians) to the very heart and most public part of the city, its market place, the agora.

A town with lots of residents cannot have every home sitting in the middle of physical concentric rings. That's a nonsense. That said, some historic cities, some Italian ones like Florence (for example), do have something akin to this structure.[4]

4 Where a city was dominated by a few very rich families, then each family tended to inhabit a quarter of the city that it might effectively call its own and the structure of the city around fulfilled the purpose of providing a high level of privacy to that rich family.

The very local should be inhabited by a very local community – perhaps some 150 or so people[5] and not many more. Local amenity, such as a park or play space, should be located and designed into the urban fabric in such a way that it is most readily available to that immediate community and less readily accessible (possibly through controlled access and certainly by design of the built form) from a wider public.

The next layer, according to sociological thinking, would be a part of the city, which is inhabited by around one or two thousand head: but not much more. Again, this links back to our roots, where it is thought that tribes frequently linked together to create super-tribes of up to (around) 10 tribal groups ($10 \times 150 = 1{,}500$) (Dunbar, 1998). This is pretty much the limit of our ability for facial recognition. We might not *know* the 1,500 members of the larger group, but we can recognise most, if not all, of them. Beyond this and people quickly become true strangers. For urban planning, the next layer of the city would be a quarter inhabited by this thousand or so heads, where amenity and other facilities (say, an early years primary school) would fall within this locality – again readily available to those within and less accessible to those from without. Reduced accessibility in this context should not be construed as being gated enclosure; rather the urban structure can be designed in such a way that anyone entering such a residential enclave feels as though they are intruding into a community's domain.

Moving beyond this scale, the direct effect of this third driver on the physical fabric of the city rapidly diminishes. This is for two reasons. On the one hand it becomes diluted and over-ridden by the city's needs in respect of the other drivers – purposeful movement and wider social/economic interaction. On the other hand, the communities within the city become more and more a morass of overlapping concentric circles: various churches, each acting as a focal point for their respective communities, and larger schools or hospitals, again each acting as focal points for their respective communities. The activities of these larger social communities will not be so readily expressed in physical form because their movement patterns are only a small part of the other daily and necessary movement patterns of the population, seeking to obtain their daily bread, do business and earn an income.

5 150 is thought to be a rather magic number in modern sociology (Dunbar, 1992; Dunbar, 1993). Research has suggested that human beings evolved to live most naturally in tribal groups of around 150, where any group growing significantly beyond 150 heads would become unstable and tend to split. It is thought that the natural community, where an individual can easily *know* all members of the community (not just name, but also kin relationships and history), is no more than 150. Much beyond this number and any community will effectively contain strangers, whom the individual may recognize but not *know*.

Balancing the Drivers

Bringing this together, we can see that the structure of the city is driven by three underlying drivers. These operate at every scale (whilst a household needs privacy, the occupants must still have good access to the town around). But the relative strength of each of these drivers, in terms of how they each influence the evolution of the cityscape, varies across different spatial and temporal scales.

The primary driver functions at the scale of the entire town or city. It is very short-term, where failure for a town or city to respond to this driver can have immediate consequences. It is this immediacy of this driver that makes it frequently take priority over the other drivers and thence to dominate the urban landscape (especially in a world with automobiles).

The second driver operates at an intermediate scale in both time and space. Meeting this driver in terms of urban structure means enabling the city to operate in economic and social terms as a coherent entity, in particular enabling business activity. In this respect it can be appreciated that inability to function effectively will not be immediately life threatening to the survival of the population of the city. But compromises in the town structure with respect to this driver do reduce its economic success in the medium and longer terms – businesses are simply less able to operate to their maximum effect in such a city.

The driver that operates over the smallest scale has an impact on the degree of success of the city over the longest time horizons – measured in decades and generations. Failure to create and build good design solutions for housing the population of the city will have greatest impact on the rearing of the next generation, affecting how one generation is able to socialise and teach the next generation the social and technical skills necessary to enable the future city to compete in an increasingly competitive inter-connected world. Given the long timescale associated with this driver, it is the one which is most easily over-ridden by other seemingly more important, more urgent, competing factors – urban communities become torn asunder by a new concrete highway or some other intervention deemed essential for the immediate shorter-term business needs of the city.

The way that a particular townscape solves the inherent tensions created by these underlying economic and social drivers determines what sort of place it is to live in and how successful it is at providing for all the needs of an inhabitant population. Places that we think of as being high quality urban environments represent those that best get the balance right. Yet the interplay and expression of these drivers and the consequential tensions can be very subtle on a day-to-day basis. The manner in which a road might divide a community, by deterring people from crossing the carriageway on a daily basis because the cars are just a little too numerous and fast, is invisible, extremely insidious and only knowable to those within the community itself. But there it is: the friendships, relationships and interactions across and along the road are just that little bit weaker or simply fail to occur, such that the community cannot genuinely be considered to be cohesive;

that essential support that neighbours can and do provide for each other in communities elsewhere simply doesn't happen. The impact is mild or non-existent on any particular day; but over two decades of a child's time spent living at home it can mean the difference between a young adult who values him or herself and participates in the surrounding community and one who doesn't.

Urban Design

As noted at the beginning of this chapter, in the last few decades the discipline of urban design has become highly scientific. In the absence of any coherent theory for the subject, urban designers see their role to know every nuance they can about the local urban landscape and natural environment and, where possible, information about the local society and community too. But then what?

At some point in the process, the urban designer can incorporate into his design solutions various rules of thumb, such as those promoted by CABE through its *By Design* publication.[6] And if the development being proposed is small enough, then that might suffice. Though even at the smaller scale, the failure to comprehend the true essence of housing has led to large quantities of exceptionally poor new residential development over several decades (in the UK).[7]

At the scale of masterplanning a new area of town, there is very little to-date within the discipline of urban design to guide the designer. As one leading architect said, "Now's the part where we play God". At the masterplan scale, once all the contextual factors – land typology, waterways, existing infrastructure, planning restrictions, etc. – have been worked around in order to minimise construction costs, then the urban designer is free to dream. More likely an urban structure will arise, which is wholly predicated on working round those contextual factors, so as to minimise a developer's costs. And if the urban designer were to suggest otherwise, he or she stands on very weak ground to argue otherwise:

> I know, Mr Developer, that the solution that I'm suggesting will cost you several million pounds more. But it just … feels … that that particular solution would produce a much better outcome, a much better quality place.

Intuition and feel do not add up to much when faced off against hard cash. Unless there is some hard science to support it.

The ideas presented so far in this book, brought together in this chapter, are a first attempt to begin to provide a coherent *theory of the town* to underpin urban planning and urban design. They are tentative proposals. But if they are proven to have merit, then hopefully it is apparent that they would greatly facilitate at

---

6 There is more discussion on By Design in Chapter 13.

7 Personal observation of the author, having worked extensively in the housing industry over the last two decades in a capacity of seeking to improve housing quality.

the very least the strategic masterplanning stages of new development projects. When combined with the ideas in the next chapter, then they will hopefully contribute towards better solutions for city-wide planning and lead to a more sustainable city.

Chapter 11
Urban Energetics

This chapter returns to the issue of urban density, focussing on the relationship between urban density and city-wide energy consumption: how does energy use by a society vary according to the population density, at which that society chooses to live? This has been a subject of hot debate over the last 20 years, in part (as will be seen) because the core academic research on the subject is logically incomplete.

Existing Research

For those who have previously explored the energy-density relationship, they will know that there is a singular piece of research, to which everyone refers. Twenty-five years ago Newman and Kenworthy (1989) conducted analysis of the energy consumption of numerous different cities around the globe and then plotted these quantities against average population density (inhabitants per hectare) for each of these major conurbations. The graph below is a replay of their frequently referenced and reproduced output. It very clearly shows that the denser the city the lower the energy consumption. It categorically demonstrates that the low density American dream performs very badly, producing major gas guzzling metropolis. Newman and Kenworthy's research cannot, however, be the whole story; their analysis pertained only to the gasoline consumption per person; it therefore ignores numerous other aspects of energy consumption within cities, which also vary with density.

Consider, for a moment, a thought experiment. Compare two extreme cities. One is a singular skyscraper – everyone living in one, very tall building, in which all aspects of the city are situated. The other is a one-road town, where every shop, home and business sits facing onto a single road, stretching for as long as the skyscraper is tall. The number of inhabitants in these two cities is identical. The movement patterns within each city are equivalent: in one case being vertical, up and down the building, in the other horizontal, along the road. And energy consumed within each shop, business premises or dwelling is equivalent in either scenario. Apart from the views from the skyscraper, a major difference between these two cities is their calculated density: the population density (people per square area of earth surface) is manifestly greater for the skyscraper.

The laws of physics (gravity) demand that the population inhabiting the skyscraper consume considerably more energy moving around than the population living along the road. In the context of gravity, vertical transport is inherently more energy intensive than horizontal transport (see Box 10.1). So, in this thought experiment, the high-density scenario has to consume more energy than the low-density city. Yet, that previous, frequently referred to research, suggests that the more dense (and high-rise) the city the less energy it will consume: a simple relationship. But this relationship must logically break down at some high density.

Figure 11.1 Newman and Kenworthy graph

Box 11.1: Vertical verses Horizontal Transport Energy Consumption

Comparing the energy consumption of vertical verses horizontal transport proves to be a difficult task. This is in part because identifying the actual energy consumption per person for elevators or lifts is itself hard. Those designing lifts rely on rules of thumb on energy consumption, which seem to have only a loose relationship to reality and can vary widely from one lift system and building height to another. In addition to this, lifts generally use quite efficient electrical motors, whereas horizontal transport is still largely by relatively inefficient petrol or diesel engines. Lifts also use counter-weights to reduce energy consumption, the weight of which is determined by the average estimated occupancy of the elevator. And lifts use electricity, where the efficiency in the consumption of electricity will itself vary significantly from country to country and city to city, depending in large part how far the city is from the power station generating the electricity.

The best figures, found by the author to-date, which provide an understandable comparison between vertical and horizontal energy consumption, suggests that the average lift journey per person per kilometre will consume around seven times the energy consumed for the same length of journey by an electric car, such as a Tesla Roadster (Fat Knowledge, 2007). Whilst there is debate on the accuracy of the figure (Ask Pablo produces a slightly lower number (Triple Pundit, 2007)) and as noted above it is difficult to state accurately how much energy lifts actually consume, this does give a sense of the order of magnitude difference between vertical and horizontal motorised movement. All the figures for lifts suggest significantly more Wh energy consumption than different modes of horizontal transport (Kenworthy, 2003). In a high-rise city where a significant number of people need to use vertical transport, then consumption of energy in transport can be expected to increase dramatically with increases in average building height.

As noted, in the original research by Kenworthy and Newman only the gasoline consumption by the relevant cities was measured; they omitted the electricity consumption. Furthermore, comparing vertical to horizontal transport is itself only part of the story, which ignores other factors such as energy consumption associated with pumping water up tall buildings, the energy consumed seeking to maintain skyscrapers and, not insignificantly, the huge initial energy cost in building very tall towers.

Deducing Energy/Density Relationship

When comparing energy consumption across different densities, the only factor that proves to be relevant turns out to be that of mechanical transport of goods and people. Both logically and also tentatively proven by research (Steemers, 2003), the energy consumption within dwellings is essentially invariant across different densities (or at the very most varies only marginally in comparison to the variance seen in transport). The remote farmhouse has no reason to consume much more or much less energy in operation than the city centre flat. Even if there are discrepancies in certain cases, at a statistical level for any society of a particular level of affluence rural dwellings and urban dwellings consume around the same amount of energy: lighting, heating, hot water, cooking, white goods and electronic equipment. In fact, it is affluence itself that is the key determining factor of absolute amount of energy consumed; for any particular society it is then fairly constant across town and country. The same applies with regard to the urban or rural supermarket or the city centre office verses that of the remote business park commercial premises (in the latter scenario, excluding movement of individuals by mechanical means around the buildings). It can, therefore, be reasonably assumed that energy consumption within buildings, excluding internal mechanical transport, will be fairly invariable across the density range.

The parameters of most interest are, therefore, the movement of people and goods around the city, horizontally and vertically. The research by Newman and Kenworthy effectively provides the relationship between energy and density for the horizontal component, albeit it excludes those elements of surface public transport that are powered by electricity. Adding in energy consumed in such public transport would marginally temper their results, suggesting that the fall in energy consumption for mechanical horizontal movement for the higher density cities would not be as great as they had calculated.

Turning to vertical transport, the very low-density city uses next to zero energy for this purpose. Increasing density, vertical movement by mechanical means begins to kick-in once the building heights start to exceed 4 storeys.[1] By the time that average building height is over 8 storeys, then by deduction at least half of all residential journeys made by individuals and goods must have a vertical mechanical component. As noted earlier, on a like-for-like basis of equivalent efficiency in motors, vertical transport is inherently much more energy intensive than horizontal transport. There is no escaping this fact. So as the density of the city rises above an *average* of 8 or so storeys, then energy consumption in vertical mechanical transport must rise very rapidly: in fact, far quicker than the graph in the other direction for horizontal transport (when density reduces).

Putting these two relationships together produces a graph along the lines of that shown in Figure 11.2. The graph in Figure 11.2 is actually inverted from the Kenworthy and Newman graph in Figure 11.1. The horizontal transport curve rises rapidly through the threshold between high-density and low-density (fastest at the 80 to 50 dwellings per hectare range); it then levels off at much lower density because the level of transport between dwellings begins to fall away and the city (as such) loses coherence; instead there exists a landscape of homesteads either operating self-sufficiently or functioning through dispersed urban centres (towns and villages).

1 The London Housing Design Guide only requires lifts in buildings above four storeys. It is assumed that people can and will readily use stairs and do not need a lift up to and including four storeys.

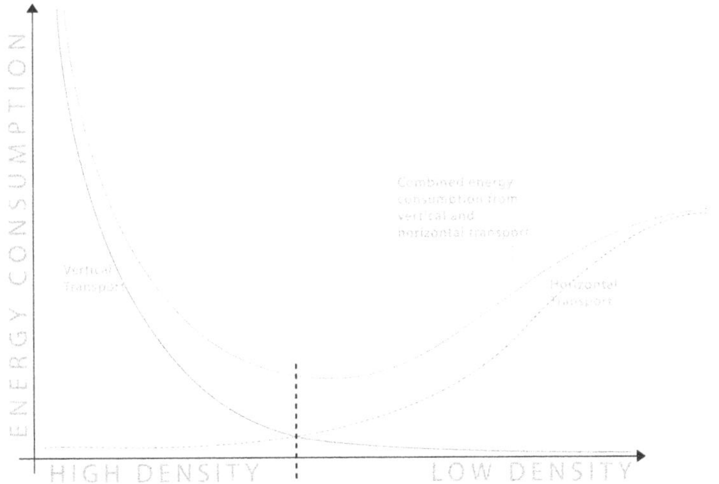

Figure 11.2 Full relationship between energy and density

Deriving the Density for Minimum Energy Consumption

The following logical arguments could easily be followed up and proven through computational analysis. This has not been done by the author on the premise that the basic reasoning itself is very strong and that history itself proves the argument (as will be discussed further, later in this chapter).

Reflecting back to Chapters 5 and 6, a reasoned argument was provided to explain how different land uses work together in the built environment. Based on economic arguments of ratios of different use types and knowledge of human being movement habits (as in, how far people will tend to walk before resorting to some form of mechanical transport), it was proposed that a stable vertical mix of uses can be sustained where average urban densities reach around 250 dwellings per hectare. At this density, there can be a ground level layer of retail and commercial uses topped by around five to six storeys of residential accommodation throughout an urban quarter. At this density, each local resident can access pretty much all his or her daily requirements from the surrounding shops and businesses without need to resort to car, bus, tram or train. In turn, there is sufficient local population within a walking catchment to sustain economically all those ground floor retail and commercial premises.

At this high, but not super-high, density, then use of mechanical surface transport is relatively minimal. Clearly there is still need for underground trains and surface trams and buses and taxis, because local people still need to travel further afield on a regular basis. And there is still a need for private transport, if for no other reason than to bring supplies into local shops and business premises and cart away waste. In respect of vertical transport, at an average of around seven storeys, a sizeable proportion of all movement up and down buildings will still be by foot (making a very broad-brush assumption that people will tend to use stairs for the first three or four floors for most journeys (as previously mentioned)).

Looking at this scenario in energy terms and considering the vertical and horizontal movement patterns by the resident population, it is at around this density that reliance on all forms of mechanised transport is minimal. By deduction it is at approximately this density that energy consumption by a city is minimal; detailed computer analysis can elicit the actual minima, but logically it will not be very far away from these density levels. Increases in density and thence average building heights from this point will invariably lead to a very rapid rise in energy consumption associated with elevators and escalators (approximating exponential increase). Reductions in density from this point lead to incremental increases in use of surface transport systems and associated growth in energy consumption associated with horizontal movement.

The mathematics is difficult to refute. A city built at somewhere in the region of 150 dwellings per hectare to 300 dwellings per hectare, averaging across a large urban area to keep building heights to a minimum, would naturally give rise to the most energy efficient urban form. This is a much higher density than many of us might choose to live in the UK and US (at least in the current car dominated world), especially those with families; it is, however, fairly common place across the European continent.

The Low Energy City

The proof, if needed, that the logic above is correct can be obtained by observing history. Those cities around the world which most closely approximate in density terms the minimum urban energy scenario derived above are those cities which evolved to that position before the automobile came into mass usage. Around Europe, many cities grew rapidly in the late nineteenth and early twentieth centuries in a world where the car did not exist or still had minimal impact on the travel movements of the majority of the population. London differed from many of these other urban centres in that it had access to copious quantities of relatively cheap coal; this enabled London to sprawl through the creation of early public transport systems, especially trams. Other European cities simply could not afford to sprawl so early. As a direct consequence, just as the structure of the natural town from time immemorial was determined by most energy efficient form, so too these modern cities quickly iterated towards a structural form, which was the most energy efficient and therefore economically cheapest to operate. Somewhere around 250 dwellings per hectare became the norm with most buildings built to around 6 to 8 storeys.

These relatively high-density urban centres manifest an intriguing dichotomy: they are both very rigid and very fluid at the same time. The physical structure of the city at this density is extremely durable and long-lasting in three respects: in terms of structural layout (street pattern), in terms of the aggregations of different land uses and in terms of the super-structure of the buildings. As explored in previous chapters, those streets, which have become, for instance, a place for eating as a public destination will tend to remain that way for decades and decades, even generations. Yet within this rigid structure, the human population flows: occupation of both dwellings and shops and business premises is perpetually on the move. A street may remain for a very long time a focal point for an aggregation of restaurants, but individual restaurants will come and go quite rapidly. And though the tides change, with the movement patterns of individual people changing week on week and year on year as people move into an area, around an area and then away again, at a statistical level movement patterns will remain highly consistent and predictable over time.

At these densities the human society and its economy is a liquid forever flowing, waxing and waning, through the very fixed physical structures of the city, making relatively little in the way of impact on those physical structures over time. In this respect, the city is both energy efficient on a day-to-day basis as well as being energy prudent in terms of changes to the physical form of the city: once infrastructure is laid down, it remains indefinitely, requiring maintenance and up-dating, but not fundamentally changing.

The Consequences of Public Transport

From the lowest energy configuration of a city public transport has allowed cities to sprawl outwards, as happened with London, creating nodes of activity at stations along linear routes leading away from the city centre. The outcome is the classic city diagram, as depicted in the Lord Rogers' Urban Task Force back in 1999 (Urban Task Force, 1999), now embedded within town planning policy such as The London Plan – see Figure 11.3.

As noted in Chapter 6, for public transport to be genuinely economically viable then it cannot operate below a certain density threshold: the rule of thumb is a minimum of around 50 dwellings per hectare. As discussed previously, this is still quite a high density, higher than most UK and US suburban densities. Furthermore, above the 80 dwellings per hectare mark, people can live out their day-to-day lives without regular resort to private vehicles; whilst they may still need to have a car for some journeys, this will be for the minority of times leaving the household.

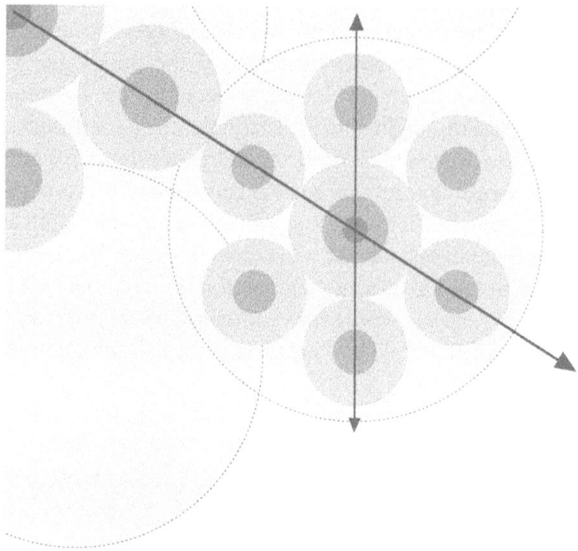

Figure 11.3 Urban density and public transport nodes

Where a population is using foot and public transport as the primary means of movement, then this transport infrastructure holds the city together. The urban landscape is forced to retain a high degree of coherence over time. In fact movement patterns are likely to be very predictable. In the case of the much higher density circumstance, considered above, a new occupant to a dwelling may have a very different routine to the previous occupant: using different shops, leisure facilities, tube stations and routes to work, as compared to the erstwhile tenant. But in the slightly lower density situation, a new occupant or family moving into an urban home in walking distance to shops and train station will most likely have very similar movement patterns to the previous occupants of that house: travelling to work by the single train station, walking to the closest shops, getting to the one school that lies within walking distance, and so on.

The overall energy consumed by this lower density urban situation may be much higher than the city centre situation, because of a much greater reliance on mechanised transport – primarily public. But at least the structure and configuration of the city remains coherent over time. The predictability of the movement patterns, which are fixed onto the landscape by the permanent public transport infrastructure, ensures that the city retains its structure for an indefinite and lengthy time period. This is not the case with what follows.

The Consequences of the Car

When the average density of the urban tableau falls below 50 dwellings per hectare, then it was shown in Chapters 4 and 5 that two things happen: different retail centres begin to compete for the same customer base and as a consequence erode each other's land values, and simultaneously there emerges a dissociation between different land uses. Below this density threshold there is no longer a geographical correlation between shops and homes; residential values can no longer be linked to accessibility to any specific urban centre; land value no longer has any link to centres of gravity of urban areas. As a consequence of these changes, the geographic landscape itself becomes fluid. At the higher densities considered above the human population is a fluid flowing through the fixed physical landscape; but at these lower densities the landscape itself becomes a constantly changing sea-scape.

The implications for energy consumption of the city are immense. Not only are these lower density urban scenarios very energy intensive to operate on a day-to-day basis, consuming vast amounts of energy with everyone driver everywhere on almost every journey, but also these urban landscapes do not retain any consistency over time. They need to be continually rebuilt. Every time there is a new dead mall, not only do the buildings themselves become redundant, but so does all of the infrastructure that supported the previously thriving mall. The need to continually re-build the whole landscape becomes incredibly sapping on the inhabitant population and economies. It is a very inconvenient truth that such urban landscapes are fundamentally unsustainable, either energetically or economically, for any sustained periods of time. The incredible waste apparent across America's urban landscape – dead malls, derelict business parks and deserted housing areas – are testament to this conclusion.

The Energy Conundrum

Fundamentally the city is not unlike any other physical system. If you *pump*[2] energy into it, it will change. Take a lump of ice and heat it (pump energy in) and it will initially become a liquid and then a gas: a highly fixed and structured system loses its integrity and becomes fluid and mobile.

In the case of the city, we see the structure of the urban landscape migrating from a low energy form, whether a small market town (with retail fixed in the centre of a residential hinterland) or a six-storey European walkable city. The graph shown earlier in this chapter suggests that, in adapting to a much higher energy flow environment, the city has two directions to travel – up and out (see

---

2 Clearly in the case of modern economies we do not 'pump' energy into the system as such. But micro-economic theory tells us that the price of energy will determine its level of usage. Where energy is cheap, then the economic system will adapt to use more of it; whereas if energy is expensive, the economic system will evolve to be more efficient.

Figure 11.4). The outcome of this is exemplified by some of the new middle-eastern cities, such as Dubai and Abu Dhabi, as well as various extremely energy hungry US cities such as Houston, Los Angeles and Detroit (see Figure 11.5). Contrasted against the low energy historic European centre with six to eight storey build height throughout, we see these new cities with incredibly tall towers surrounded by a huge expanse of very low density dispersed city. The intermediate (yet relatively high) densities, the midtown, associated with the many European capitals (Paris, Madrid, Berlin, etc.) simply do not exist. The low energy urban solution cannot, in fact, exist because it simply does not consume energy at the rate at which it is being *pumped in.*

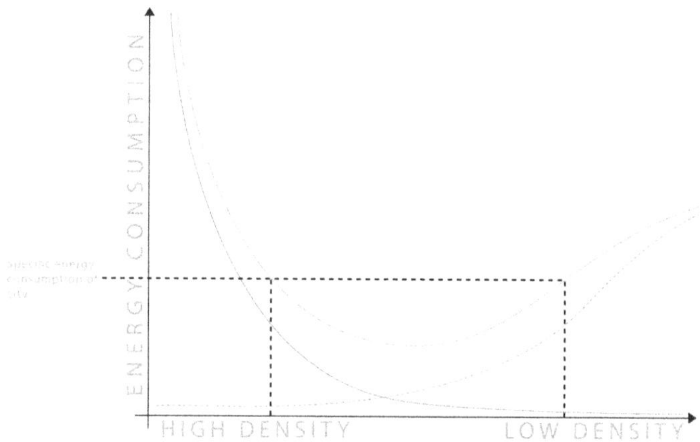

Figure 11.4 Urban form arising out of specific high level of energy input

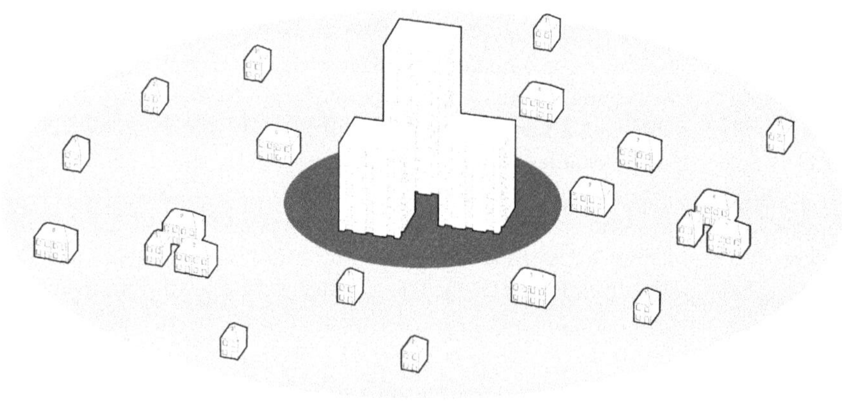

Figure 11.5 Typical urban form of high energy cities

In a world where there is so much verbage about embracing sustainability and seeking to address global warming, these are uncomfortable truths. The simple fact is: around the globe we are currently building cities in a form, which responds to copious quantities of cheap oil. When this fuel runs dry, unless there is some equivalently affordable replacement, these cities will simply not be able to exist in their original built forms. The suburban sprawl can quite rapidly change; but what will happen to those looming skyscrapers is anyone's guess.

The final chapters in this book turn to the political, practical and then social implications of designing our cities at different urban densities and, in particular, the contrast between the moderately high density (above 80 dwellings per hectare) density and low density (below 50 dwellings per hectare).

Chapter 12
Urban Development and Urban Planning

Throughout this book an attempt has been made to consider the way that cities both operate at a point in time and how they naturally evolve over time. These are both critical issues to understand for anyone seeking to carry out either development activity within the urban landscape or planning for change of that landscape. It has hopefully become apparent that the shape of the urban landscape is generated by some quite understandable economic and social forces. It has been postulated that there are three main underlying drivers that influence the shape of the city; these operate at all scales; these and the tensions between these drivers represent the urban dynamics that affect the form of the super-structure of the city all the way down to the way that we design the inside of our homes. Following on from this, it has been set out that cities adapt according to the energy availability, expanding upwards and outwards when more energy is *pumped* into them. Cities are living systems, evolving their form according to their economic and environmental context and, critically, energy availability. And the manner in which they evolve has major implications for the quality of life and lifestyles that members of a city's population can access and lead.

Drawing on this better understanding of urban dynamics, this penultimate chapter provides some food for thought for both *developers*, those who seek to gain economically out of changes in the urban landscape, and *planners*, those who seek to manage and control that urban landscape for the public good. These players are the yin and yang of change in our townscapes, both essential to allow change to happen, to allow the city to evolve.

Urban Development – Low Density – below 50 Dwellings per hectare

The manner in which a developer should view the city should be very different according to the existing urban density. Where urban densities are low and the land use planning system is very liberal, then any investor in the urban landscape must be acutely aware that everything is necessarily short-term. At low densities, the urban landscape is in continual and unpredictable flux. The aim for any private developer is to ride that flux, surf a wave here, surf a wave there, but never to stay very long. As noted, this only applies when the planning system is very liberal.

The optimal strategy for a developer in this environment is to create a big splash, by changing the travel patterns of as many people as possible, so as to reap in a quick and large profit. Where a developer is able to affect population movement patterns, then, through the flocking that he generates, he is essentially creating a super-profit for himself. Land, which was heretofore only valued as farmland, has suddenly become a destination on the urban map and accrues a much higher value accordingly – the profit can be very great indeed.

Whilst a super-profit can be achieved in the short-term, there is, however, no certainty that the land value will endure. To this end, all new developments have maximum value at the moment they open. Thereafter, they are on a decaying trajectory as other newer and fresher competition appears on the landscape and the focus of attention of the car-mobile population moves on. The flock goes elsewhere; see *The Economist's* essay on A Planet of Suburbs for more discussion on this point (*The Economist*, 2014).

This short-termism applies no matter the land use proposed by the developer – leisure, hotel, retail, commercial or residential. At low densities, there is no logic to the landscape – any land use can site itself anywhere, at the whim of private enterprise.

Where there is a planning system in place, then in the low-density context it is that planning system which dictates categorically what type of land use can go where. As explained in earlier chapters, below a critical threshold all land uses dissociate and no longer have any meaningful geographic relationship. Land has no innate value defined by its spatial proximity to anything else on the landscape. Land value is therefore determined purely by the planning system – what the planners dictate is an acceptable land use at any particular spot becomes the 'law of the land'.

This puts developers and the planners head-to-head in an uncompromising battle.

Where there is a relatively inflexible planning system in place, such as we have in the UK, then the upfront risks for a developer can be considerably higher – trying to gain planning permission. But conversely long-term profitability is also much, much greater. The risks faced by any one developer are also higher for his other competitors. If a developer does succeed in gaining planning permission for something that might affect population movement patterns, and thereby generate a super-profit, then his investment is secure for much, much longer. The fight is worth it, because profit is assured both short and long term. The landscape thereby becomes a fierce battle-ground, where the developers desperately seek out the potential gold-mine (a shopping centre in the middle of nowhere), yet find themselves continually thwarted by the planners.

In terms of design and construction, a more rigid planning system should in theory change a developers approach to investment in the built form. The retail centre is likely to be profitable for a much longer time period, so it may be worth building to a higher specification. But perhaps because of the inherent uncertainty, that land value is dictated by (what are in essence) political whims, a short-term attitude remains amongst any developers – housebuilder or retail developer or other – who are constructing in out-of-town locations. And so it falls to the planning system to enforce any design and construction quality. Again, this places the low-density suburban and rural developers at complete logger-heads with local planners.

It comes as no surprise then that the suburban and rural housebuilders in the UK perpetually lobby central Government to beat up the planners and the planning system. Development and town planning in low-density contexts are inherently antagonistic, even hostile, and no amount of diplomacy can avoid the perpetual war that is consequently and consistently waged.

Urban Development – High Density – above 80 Dwellings per hectare

The fundamental difference for any high-density development is that the landscape is fixed. All urban land has innate value, which is determined by its relationship to human activity in the surrounding townscape. Movement patterns are far more fixed into regular routines and channelled through public transport infrastructure; it is much more difficult to influence sufficient numbers of people to travel to new destinations on the map. Existing urban centres and public destinations maintain their credibility, enduring as centres of human activity over time. Development activity consequently takes a very different form. Instead of seeking a super-profit from changing in any significant way the underlying value of a piece of land, the urban developer must look to profit in other ways. Key strategies include:

- Land Assembly: marrying together neighbouring plots of land in order to create something larger and better than could previously be sited at such location – the whole value is more than the sum of the parts.
- Design: changing the nature and character of a location through good design, so as to make an area more attractive.
- Land banking: obtaining land in decayed parts of the city and taking the long view, waiting until it starts to regenerate and/or gentrify.
- Land use change: following the market by obtaining buildings of one land use and switching them to another, which has recently become more valuable through increased demand.

Every now and then an opportunity presents itself for a developer to make a big splash in a highly urban context, perhaps obtaining derelict land next to a potential or active transport hub, or when a major new transport interchange is created. These are, however, very much abnormal opportunities. In the last decade there has been a spate of such opportunities in London (and across the UK). But these have been rapidly mopped up by eager development firms keen to harness potential super-profits in busy locations. Over the coming decades, such major regeneration projects will inevitably become fewer and farther between, or at the least more complex and difficult to realise.

In the high-density context, because of the fixes created by the urban landscape and the supporting infrastructure, land value endures. The name of the game for development activity is to have an eye on the longer term. This in turn means that developers are willing to invest much more in design and construction quality. In this context, quality can essentially be defined simply as: higher quality equals lower maintenance costs. Certainty of underlying land value gives the developer the confidence to invest much more upfront in order to construct an asset that is more cost effective to manage and maintain in the long-term, allowing maximum yield from the rents.

In this context, it can be quickly appreciated that the relationship between developer and planner is very different. It is much more of a partnership, even synergistic, looking together how to enhance the built landscape over the long-term. It is unsurprising therefore that in London and other city centres, the urban developers tend to treat the planners and the planning system with much greater respect and are much more likely to work collaboratively to achieve urban regeneration.

Planning Determination

It should by now be apparent that the role played by planning determination must necessarily be very different between the low-density and high-density contexts.

In the low-density situation it matters far less what a building looks like or whether it might have any impact on neighbours. Such considerations are essentially irrelevant. New buildings or extensions may not even be visible from the public realm and are therefore only a private matter for the landowner. Rather the key factors that any planner, thinking about the public good, should be concerning him/herself with are the quantum of development and the proposed land use. Whilst a new development may not in itself have any direct impact on the public in terms of visual impact, noise or other potential nuisances, the travel patterns that it generates are a concern both in terms of energy consumed and busy-ness of the public highway network and resultant impact on the ability of other people to travel where and how they wish.

As set out in Chapter 1, different land uses experience very different visitation characteristics, where shops and houses sit on opposite extremes of a spectrum. One key question for the planner is therefore whether the existing road network can accommodate the new travel patterns associated with such new development. And in the context of a potentially warming world, should such new development be allowed if it blatantly requires the consumption of significantly more energy?

A town planner working for the public good needs to be wary of the developer looking for a super-profit by seeking to change significantly travel patterns in the area, regardless of whether the road network is up to the task or might be adapted accordingly. Such super-profit is indicative that either new infrastructure is required or that existing infrastructure will become redundant. If a developer wishes to proceed, he should be contributing investment in the infrastructure. Essentially the existence of super-profit is suggestive that a new development is creating economic, social and environmental externalities, which need somehow to be mitigated, or giving rise to unnecessary physical competition on the urban tableau and thereby eroding land value elsewhere.

In the high-density context, the situation is the reverse. The considerations of main concern are no longer land use or quantum of development, rather the town planner needs to pay most attention to the aesthetics of the new development and the potential impact on its immediate neighbours – by definition, at higher density neighbours will be more easily impacted. As considered in previous chapters, investment in the built environment at higher densities has the opposite effect to the notion of externalities: land uses cooperate to build land value. So, well thought out investment in new buildings will help to raise the quality of an area and benefit the surrounding townscape and other local land owners.

At high densities the overall quantum of development proposed in any particular location should not be such a great concern to the planner (with the exception of particularly larger investments), because for the most part a developer will be constrained by the surrounding landscape in two respects. Firstly, without being able to have any significant influence on movement patterns, there is no economic benefit for a developer to propose, all of a sudden, double or triple the amount of floorspace (at least for retail or commercial purposes); it would simply not be economically viable unless something was happening to change the level of accessibility of the location. Secondly, where land is highly bounded by neighbours, it is the potential impact on those neighbours, which will limit what can be accommodated on any particular plot of land. So, in a sense, other factors constrain quantum, without the planning system needing to go beyond ensuring that those factors are properly being taken into account.

At higher densities, a system for extracting planning gain from new developments is still required, but should operate in a different way and the finances raised should be applied differently. Firstly, it is clearly required to offset any impact in the immediate vicinity either by compensating individual neighbours or recompensing the local community in some way. And secondly, and only where there is an up-lift in quantum, planning gain is required for investment into the increased stress on infrastructure. At the higher densities, in an ideal situation, such tax would only apply for the additional accommodation, over and above what already existed in that location.

In summary, at low densities the land use planning system needs to focus on quantum and land use type, whereas at high densities the considerations in the determination of applications should be much more on issues such as neighbour impact and aesthetics (the visible affect of the new development on the public realm). There is clearly a spectrum between these two extremes, albeit as noted there is a fairly well-defined threshold below and above which these different influences need to be addressed. Interestingly, the practice of determination of planning applications in the UK actually fits the model described above fairly well. Generally speaking, developments in the centre of London go through far more rigorous assessment in terms of aesthetics, environmental assessment and neighbour impact than do those on the edge of or out of town. It would be good, however, if planning theory as taught at university were to catch up with this practical reality.

Chapter 13
Urban Design in Practice

The purpose of this chapter is to spell out how the theoretical ideas presented through this book are connected with urban design as practiced. In the UK there are, perhaps, three bibles to which urban designers revert for guidance on the practice of urban design: these are *By Design* published by the Commission for Architecture and the Built Environment (2000) and the *Urban Design Compendium* published originally by English Partnerships (2000) (and somewhat later followed by a Part 2 (English Partnership, 2007). Between them these three documents fairly comprehensively cover the territory generally understood to be the remit of urban design; they set out what is considered to be best practice in both how to go about urban design and what urban designers should be aiming to achieve; with respect to the latter, in the final chapter some further discussion is provided on the origins of this guidance. Here follows, however, a quick run through of the key tenets within that guidance and how it relates to the theory set out in preceding chapters.

There have been numerous other publications over the last two decades that seek to define urban design and set out exactly what is its purpose. This will not be repeated here. The substance of this book hopefully speaks for itself and should, in due course, help to inform that debate. It is, however, worth reflecting on the context within which urban design as a profession has emerged. For much of the latter half of the twentieth century urban planning has had a core focus to respond to an increasingly car-dominated urban environment. Ideas such as the Gravity Theory, helping to inform planners what size roads should be built to connect town A to town B, combined with the practice of zonal planning (as discussed in Chapter 5) both fall within this framework. Urban design as a subject of academic and professional interest and in its various forms (for example, new urbanism) can in part be construed to be a counter-response to these previous styles of urban planning; urban design as practiced represents an evolution from urban planning where it has become clear that previous ideas either were not working or were manifestly inappropriate for higher density urban interventions and regeneration projects.

In being a counter-response to the errors of urban planning, there is a risk that the urban design guidance over-steps the mark. This is most apparent in the complete refusal to engage with the fact of urban density within the various elements of guidance. By way of example, the Urban Design Compendium does give a cursory nod towards urban density, but it does not explain how the remainder of the guidance within its pages should be adapted for different urban densities. There is a strong inference that all aspects of the guidance should apply to all levels of urban density, without any recognition as to how the urban landscape changes at different levels of intensity of human activity (as spelt out in this book here). Furthermore, the manner in which the *Urban Design Compendium* discusses urban density bolsters the appreciation that it is seeking to counter previous zoning practices, where urban density might be pre-defined across a whole area. There is a consequential risk that all that such urban design guidance does is create slightly better designed places, but which are nevertheless still enclaves or zoned locations, which are still connected by roads and effectively represent private destinations on the urban map (see back of Chapters 3 and 4 for the introduction to the concepts of private and public destinations).

It is hopefully expressly clear from the discussion through this book that what is appropriate in various locations will be fundamentally different depending on the intended level of human occupation and intensity of activity that is desired. In practice not everywhere can be a *place*. There is always a need for major access routes to service towns and cities; and the larger those towns and cities are the larger those access routes; and the larger those access routes the more those roads will become divorced from their immediate surroundings and create hostile edge conditions. And where such access routes cross, then it is inevitable that such locations are by their nature transitory and not for lingering in; they become *non-places*. Furthermore, trying artificially to activate the edges of these major access routes is a futile endeavour; the edges of such roads through the town will always be uncomfortable environments (especially those urban motorways that are essential in major cities). You cannot create *places* along these roads unless you intend to tame the road; but before you do that you need to consider carefully how fast and furious the traffic speeding along that road must run (or not) to service the daily needs of the city.

Separation may be appropriate (if you can afford it) for the very largest urban motorways – raising them up or burying them. But we have discovered from the errors of the 1960s and 1970s that it is not desirable to seek separation on other lesser roads (for instance, the various ring roads that now throttle cities such as Birmingham and Coventry). Nor is separation of traffic from pedestrian desirable at the very local level – witness those failed estates with undercroft car parking and high level pedestrian walkways.

It is instead hopefully clear from earlier chapters that the strategic objective of urban planning and urban design needs to be a balancing of those competing tensions between macroscopic accessibility to a town or city centre and local accessibility to the residential parts of a town – generally, for most towns, without separation between road traffic and pedestrians. In this respect, existing urban design guidance tends to focus primarily on ensuring that new schemes are properly stitched into the existing urban fabric and street patterns, generally seeking to maximise the number of connections between the existing urban pattern and the new. That is all very well, but if that leads to the creation of rat-runs and through-routes, which would detract from the types of streets and places that the designer is trying to create or damages an existing well-functioning street or place, then such knitting could be as deleterious as it is thought to be beneficial. The thinking through this book suggests that a more sophisticated approach is required to think through thoroughly how a new piece of the city might be used – intentionally and opportunistically – by the inhabitant population, close and far. Those locations that the urban designer wishes to become true places will only do so if they can come to be genuine *pauses* within the city fabric.

The theory throughout this book is very much focussed on the functionality of the city, in particular the ground plain structure of the urban environment. Some of the thinking may be extrapolated upwards, but this has not been addressed here. In this respect the thinking in this book does not extend or amend existing guidance on matters such as appearance, scale, height of buildings, enclosure and variation in built form. However, reflecting back to the question of urban density, a cautionary note is suggested to think carefully about how such guidance is or is not appropriate or even relevant at different levels of urban intensity.

The main other areas of urban design guidance correspond to the following areas: accessibility and permeability, response to topology and landscape, mixing of uses, adaptability (also referred to as flexibility or design for change) and intentional public spaces (hard or soft landscaped). These are each considered further below. This book also contributes completely new thinking that goes beyond these existing areas of guidance: particular items of note include making explicit the difference between public and private destinations and a much better understanding of the relationship between what happens on the street and interaction with the building edges along it.

Accessibility and Permeability

Throughout the guidance that exists, there is a general promotion of accessibility and to maximise permeability through new development schemes. Cul-de-sacs are, for example, normally eschewed. With the caveat regarding the potential to create unintended roads through a scheme, balanced against the need to promote connectivity across the wider urban fabric, such guidance is correct at the strategic level. As noted in Chapter 10 on a Theory of the Town, there are three competing tensions that need balancing across the urban fabric. Two of these seek to improve accessibility, albeit at different scales; but the third tension is very much focussed on reducing accessibility, to provide privacy and seclusion at a very local level. Unthinking pursuit of maximum accessibility has the potential of replicating some of the problems manifest in many of the 1960s and 1970s housing estates, which suffer over permeability.

In this respect, it needs to be recognised that very short cul-de-sacs and other types of dead-ends are not only quite acceptable but actually very desirable, so long as they serve only small communities. As discussed in Chapter 9, tree-like urban structures, which effectively represent exceptionally large or long cul-de-sacs, do not provide good places for people to live. It should be remembered that corridors within buildings are just as much urban dead-ends as conventional cul-de-sacs; in respect of the latter, the London Housing Design Guide has now, fortunately, put a stop to the housebuilder excesses by limiting building efficiencies and preventing further proliferation of blocks of flats with minimal cores and very long internal corridors – a high-density variation of the tree-like flawed solution.

So, in summary, where dead-ends can create small communities and promote use of the highway as public open space, then they are beneficial and should not be avoided simply to maximise accessibility.

Response to Topology and Landscape

Here is an area where there is clearly major confusion as a result of refusing to acknowledge the degree to which urban form should change according to urban density. The tenet of the guidance is that urban designers should seek to work with the natural landscape and topology of a site. This guidance seems, however, to be somewhat flawed or incomplete in that it fails to acknowledge the inevitable opposing tensions. Working with the topology of a site is likely to minimise capital costs to a client, for example running roads and servicing along natural contours and may improve values by harnessing views and sightlines from individual properties. This approach may be acceptable at lower densities, where the highways are primarily roads and the alignment is irrelevant to the wider urban fabric (remember, roads are single dimensional objects that can weave and wander randomly across the landscape, so long as they connect A to B). But at higher densities, where there are streets, or an attempt to create streets, then urban structure is much more critical and should be considered as possibly more important than simply flowing with the landscape.

In considering this question, one has to ask oneself how San Francisco would look, had it been designed according to modern guidance of urban design. It is somewhat unlikely that those highly dramatic streets running up and down steep hillsides would have been considered appropriate or desirable. Yet they provide some of the most sought after, desirable and most visited parts of that city.

Response to typology is consequentially a matter that needs to be considered in the context of the density of the urban environment being created. At high density there may well be other factors that should be prioritised over seeking to flow with the natural landscape and considerations of minimising capital costs. Essentially, once you are in the heart of the city, the environment becomes so unnatural (i.e. entirely human created) that social and economic requirements for a better functioning urban fabric should be seen to dominate over the natural environment.

Mixing of Uses

And mixing uses is another area where guidance should vary significantly according to the extant and proposed urban density. Within existing guidance there is a complete failure to acknowledge the 50 to 80 dwellings per hectare magic threshold and how the urban landscape operates above and below these densities. This can lead to serious flaws in schemes, more often than not aiming to achieve a level of mixing of uses that simply will not sustain itself over time. Even at relatively high densities, relative to what we are used to in the UK, vertical mixing of uses remains economically unsustainable and horizontal mixing needs to be careful considered. Much more realism is required within urban design guidance on this matter, recognising how land uses naturally vary across the urban landscape, what ratios to expect and how uses of a similar type naturally aggregate together up to very high levels of footfall and intensity of human activity.

Adaptability

This is one of those motherhood and apple pie objectives. Of course it would be good to provide flexible and adaptable urban forms that can stand the test of time and do not need to be demolished inside of 40 years. The best way that this can be achieved is to understand the natural economic forces and urban dynamics that we are playing with as we make changes to the urban landscape – as have been set out through this book. In particular, it is important to recognise how the inherent social and economic tensions play out across the urban environment and achieve a degree of balance in any particular location. The reality is that the overall structure of the city will change very slowly, if at all; the need to access the heart of the city verses the need for local accessibility will not fundamentally change over any time horizon that we might concern ourselves with. If new schemes can recognise and work with these accessibility needs, then they will likely endure, at least at a structural level.

Even if it has not understood why, existing guidance that promotes grid patterns is at least heading in the right direction. Hopefully the ideas presented in this book provide a deeper understanding as to why the gridiron works so well and when and how it might at times be broken.

Intentional Public Spaces (Hard or Soft Landscaped)

The preceding chapters of this book have not dwelt on urban interventions such as the creation of parks and green spaces within the urban fabric. However, the same principles that have been derived earlier can be applied to decisions about where to locate green spaces and how to incorporate them.

Firstly, it must be emphasised that the creation of specific public spaces, whether hard surfaced or green, should not be at the expense of providing proper streets. With highways entirely hijacked by motorised transport and a loss of the ability of streets to act genuinely as public spaces, there has been a tendency to seek to compensate through the creation of specific, dedicated green spaces and play spaces – a playspace here, a pocket of greenery there. There is a risk, however, that such interventions end up simply being other *private* destinations within the urban fabric, even if they are publicly owned and managed, and not provide genuine additional amenity to a locality. To be successful, careful consideration needs to be given as to how any specifically created public space integrates with the roads and streets within a local area; success can be considered here in two respects, where one does not necessarily mean the other: on the one hand, does it become well-used by residents (from near and far, depending upon size and intention of public space) and, on the other hand, does it actually contribute improved amenity to the people living around or near it?

In the context of the discussions earlier in this book, there could be construed to be two principal types of public open space: incidental public space, which is intended to improve local amenity and is there for the immediate locality, and destinational public space, which is provided as an amenity for a wider population where the majority of visitors will be journeying from afar.

Taking the latter first, because it is much simpler to deal with, destinational public space is what it says on the tin. It is primarily those larger parks that are designed to provide a little bit of countryside within the urban fabric. It is, to all intents and purposes, a leisure and recreation destination. Whilst it can provide very dramatic edge conditions (witness Hyde Park, London and Central Park, New York) and provide some breathing space within an otherwise intense, high density urban environment, it does not need to knit into the streetscape; it can quite reasonably be surrounded by roads; it needs, after all, to be accessed from further afield. Other public spaces that fulfil a similar role, but which are not green parks, include those central city squares, which become destinations and congregation points for tourists and visitors from afar, usually as a result of the function and design of the buildings surrounding them, or become foci for public demonstrations: examples include Trafalgar Square (London), St. Mark's Square (Venice), Louvre (Paris) and Zocalo (Mexico City).

Within the category of incidental public space, there are two distinctly different contexts: the residential neighbourhood and the town centre. The purposes of such spaces and the principles, in terms of relationship to streetscape, are the same, but the design and character of such spaces will be quite different.

Reflecting back to Chapter 9, the context for the residential public space is that this begins to become important as packing of houses starts to impinge significantly on private external amenity for the surrounding dwellings; at very low densities, then such public space is not required, because back-gardens are sufficiently large; as those gardens are squeezed, then a compensatory level of public space needs to emerge in order to provide adequate doorstep external amenity for the local inhabitants. By far the best way to achieve such public space is by means of taming the highway to make it a true street that the local residents can appropriate for uses other than just driving and parking cars: childrens' play, socialising, parties and the like, and generally designed to generate interaction between the surrounding dwellings. This suggests that any incidental extra public space should, as far as possible, be completely incorporated into the street – it should be very much a part of the street and it should be right there on the doorstep.

The next factor that needs to be taken into account within the masterplanning of local residential public spaces is the need for the residential area to be designed according to layers of privacy; what size of community is the public space in question due to serve and how can it be located and configured in such a way that that community take ownership of it and feel that it is theirs? Far too frequently such local public spaces are used instead to act as buffer zones between neighbourhoods or thought of as places to enable different communities to interact (perhaps the new residential community with the existing, long-established residents). According to the logic set out in earlier chapters, this is a mistake; anecdotally, there are numerous incidents where this has simply led to vandalism and graffiti and under-use of the public space, because it becomes a no-man's land. At higher densities, it is buildings that should be used to demarcate boundaries, of any type, not open spaces.

Turning to the city centre incidental public space, these too need to be carefully embedded within the streetscape. They are, however, amenity space of a more transient nature and, in contrast to the residential setting, are there to enable mixing of communities from the town; they are genuinely public. Their purpose is to relieve pressure on public streets: think in terms of how modern retail developers design their Malls, locating the food court above the main cross-streets. Such incidental city centre spaces can either be provided by means of the streets themselves, consider the natural shape of the cross-streets discussed in Chapter 3 (the circus), or by being easily accessible from such *places*, previously defined as being those pressure points within the townscape.

Reflecting back to the guidance provided in the *Urban Design Compendium*, the closest that this guidance currently comes to any of the above considerations is to point to there naturally being a hierarchy of public spaces across the city. Otherwise, this guidance stresses the importance of good design and directs designers to using existing resources (such as an existing copse of trees or water feature), which is already considered above under response to topology may in high-density contexts be completely the wrong thing to do.

Conclusion

The purpose of writing this book has been to seek to fill in some of the gaps within existing urban planning and design practice, in particular by providing some better underlying theory. As will be clear from this chapter, whilst there are some good documents providing guidance for urban designers, setting out a large number of rules of thumb on how to operate and what to seek to achieve, these do not appear to be based on any rigorous analysis of why the town operates in the way that it does. Hopefully the discourse through this book can help to amend that state of affairs.

Chapter 14
Urban Design and Society

In the preceding chapters it has been made clear that the invention of the motorcar has had a huge impact on the way our cities operate. From an individual perspective the automobile has been a fantastic liberator, allowing each and every one of us at all layers of society to experience a far larger world on a day-to-day basis than could any of our ancestors, even the richest and most powerful. But with this amazing technology has come social and economic repercussions, not to mention the obvious environmental impacts (noise and poor air quality). Even if we were to invent a highly energy efficient, pollution free form of private motorised transport, this would not necessarily (at least in social terms) deliver the panacea that many seek.

In various places through this book, it has been commented that there is a severe lack of theory to support the disciplines of urban planning and design. There is certainly nothing, which might be construed as comprehensive or which links together the various disparate areas of thinking and activity. Instead, the closest that urban designers or local urban planners come to any theoretical framework at present is theory about how to go about urban design (as noted at the start of Chapter 10). This is arguably pretence at being scientific about the subject. The student of urban design is taught that he must diligently investigate and gather together as much information about the context of a new development (large or small) as possible. But once he or she is overloaded with huge quantities of information about the environmental, economic, cultural, community, legal, planning and other contexts of a new proposed piece of the city, then what?

There exist to-date three principle areas of urban design thinking, which can be used to assist an urban designer to come to a solution to a particular problem. These are (1) rules of thumb taken from historic cities, (2) learning from past mistakes and (3) geometric algorithms.

The rules of thumb pertain to those ratios of building height to street width, and other relationships, which are deemed to work well; these are included in the *Urban Design Compendium*, discussed in the last chapter (English Partnerships, 2000; 2007). Such dimensions and relationships do not have any connection to human ergodynamics or social interaction; they are taken from historic towns, especially Italian renaissance cities, where certain streets and piazzas are deemed to *feel* right. The solutions of the past masters have then been adopted as rules to be applied elsewhere. But can they be properly extrapolated to the modern context, to the modern city, which has amongst other things to accommodate the automobile? There is little by way of scientific rigour to these rules; rather they emerge from consensus amongst those who have visited and toured round our historic urban centres. This is not to say that such rules of thumb are wrong. It is just a question of whether they really have a place in our design catalogue until someone has asked why.

Our knowledge gleamed from learning from past mistakes is an extremely important component of the urban designer's repertoire. Ideas such as defensible space fall into this category. The combined works of Jane Jacobs, Oscar Newman and Alice Coleman (and any academic debate and critiques provoked thereupon) provide the major component of this area of learning. Add the insights of Christopher Alexander to the above list of urban philosophers and the library is almost complete. Like the rules of thumb noted above, this area of knowledge draws from experience, seeing how urban areas do or do not work when inhabited by real people, families, children, communities and car drivers. They provide the urban designer with a counter-play of things to avoid to run alongside the rules of thumb to follow. But they do not provide for anything that could be construed as a coherent theoretical framework – *a theory of the town.*

The final area, using geometric algorithms to predict the relative accessibility through the urban grain, is generally known as space syntax; this was introduced in Chapter 10. It is the closest that exists to-date, which could be described as a genuine theoretical framework. And while it seems to work well, to a degree, perhaps the reason why it has not yet become as widely used as its protagonists like is because it is too theoretical. Advocates of space syntax are able to demonstrate that their approach maps against the real life of functional cities, but they struggle to explain how the mathematics relates to urban form and human activity. The theory seems to work, but why? A tentative explanation is provided in Box 14.1

Box 14.1: Relationship between Space Syntax and Urban Vectors

For most people coming into contact with space syntax analysis, it comes across as a bit of a black art. Various computer algorithms are used to produce pretty, colourful maps of the town, which apparently show across a spectrum how accessible each highway segment is to all other highways in the town. Space syntax analysts claim to achieve around or over 80 per cent correlation between their results and how the inhabitant population can be seen to make use of the extant highway network across a town. In this respect, those highways deemed more accessible are found to be those that are more heavily used (especially by pedestrians). But what is this analysis really telling us about a town?

The simplest space syntax algorithms take in turn each highway segment in a town and step-by-step find the simplest way to travel from that segment one-by-one to every other highway segment in the town. A record is made of each journey, identifying all the intervening segments that are required to proceed from each A to each B. Slowly (depending on the computer in question) a picture builds up to show which segments are required more frequently than others when adding together all the imaginary journeys.

Inevitably the high street in the centre of town – at least in any well-functioning town – emerges as the most essential segment of highway through which the most number of these theoretical journeys must be made to travel from any one part of the town to any other. By definition, any journey crossing the centre of town will tend to use some part of the high street.

It is useful to look in a little more detail at how this analysis operates. For each segment-to-segment theoretical journey across a town, an imaginary (as the crow flies) line is identified connecting mid-point of a segment to the mid-point of another segment, and then the algorithm analyses how to travel along this line by means of the actual highway network of the town by means of the least number of switches in direction or turns. The premise is that when people walk from A to B, they tend to do so by the most direct route possible and the least number of changes in direction.

If you think about it, unless a particular highway segment sits exactly on and is precisely parallel to this crow-flies line, then the subsequent deduced journey through the highway network will involve crossing all those segments that are travelled along. Every time that the high street, by way of example, is used as part of a route to get from A to B, such journey almost inevitably involves traversing the high street (as well as travelling along it), unless the start point and end point of such journey is directly along the alignment of the high street.

As the picture of the town builds up, then what the space syntax algorithms are doing is identifying which highway segments in the town are most traversed. Any segment that is a cul-de-sac will automatically, therefore, come across as being of low accessibility because it cannot be traversed in making any other journeys; it will only ever be used for journeys directly to or from it and those journeys will have to be along it and not across it. But those highways that space syntax analysts claim to be most accessible are, in actual fact, simply those that are most often traversed in the panoply of theoretical journeys that the algorithms have assessed.

What becomes apparent, when considered in the context of the ideas presented in earlier chapters in this book, is that the analysts are, perhaps unwittingly, calculating rather crudely what the fractal dimension is for each and every highway segment across a town. In other words they are deducing the degree to which any highway segment is being treated by the inhabitant population as a street (closer to being a two-dimensional object and therefore representing public space) as opposed to a road (closer to being a one-dimensional object – dominated by through traffic).

Picking up on the discussion in Chapters 7 through 9, it is also apparent that the space syntax analysis misses out on the very local scale. As a general rule, space syntax experts will advise against cul-de-sacs. In their view dead-ends are bad news because they are deemed inaccessible within the highway network. Yet the discussion earlier in this book asserts that cul-de-sacs have a role, especially in purely residential parts of the town. The reason for this disparity is that, somewhat ironically, though the analysis ends up identifying those highway segments in the town that are more street-like in character, the starting assumption for the space syntax work is that every highway segment is purely a road for travelling along.

In this respect, the space syntax analysis is only considering the macroscopic movement across the town, journeys relating to Drivers 1 and 2 as set out in Chapter 10. And as specified in that Chapter, Drivers 1 and 2 are all about maximizing accessibility. The space syntax analysis ignores the very localized activity relating to Driver 3 (for example journeys between neighbours within a street), which opposes macroscopic accessibility and seeks to create seclusion and privacy within a neighbourhood.

Practitioners need, therefore, to be aware of the limitations of space syntax analysis, especially at the very local level. In contrast to assertions of space syntax experts, as asserted by every estate agent, reduced accessibility at the very localized level is actually good and should be used to help promote the creation of public space in streets in residential areas.

What seems quite incredible from the perspective of someone coming to town planning and urban design afresh with a harder sciences background (as has the author) is the seemingly complete lack of anyone asking 'why?' Jane Jacobs had a go, albeit she rapidly drifted away from her architectural roots and the physical form of cities, focussing more and more in later works on economics and cultural matters. Oscar Newman, Alice Coleman and Christopher Alexander each attempted why questions, but each within very closed and tightly defined parameters and contexts.

If one were asked to study an ants nest and explain its structural form, then it would be manifestly inadequate not to see the structure of the nest as the consequence of the actions of the multitude of ants and the activities of those ants. In the natural world context we can all appreciate very readily that form follows function. It seems bizarre, then, not to see cities in the same light. In this book I have attempted to do just that, to seek to explain the structural form of the city as something that arises out of the innumerable day-to-day activities and interactions of its inhabitants. There are clearly many, many exceptions to the rules, which have been suggested. But at a statistical level, in being human and being biological creatures, people live relatively predictable lives – eating, sleeping, earning an income, rearing children, etc. This leads to patterns of activity and motion, which again are quite predictable, wherever in the world you may be. And if you stand back and endeavour to see the long view, how a city's form develops and evolves over decades and centuries, then it doesn't take long to realise how the form follows function.

Much, possibly the vast majority, of urban design thinking to-date seems to approach urban design with the eyes of an architect. The physical structure of the city is seen as the fixed element, around which and through which the inhabitants of the city live out their lives. And it is true that once a piece of city has been constructed, we each are forced to bear the consequences of previous design decisions – for a period of time. If, for whatever reason, a designer has chosen to put an entrance to a new office block in a location, which does not happen to be a natural desire line, then anyone entering or leaving said building must walk those extra few yards accordingly. Or if in a park, instead of using a useless footpath to nowhere, cut across the grass to create a new muddy track. But such observation of the city is rather short-sighted. Any buildings or objects, which are too much of an obstruction or which are manifestly not fit for purpose, eventually come down and make way for new urban layouts and physical structures. It may take time, many centuries sometimes, and society may well itself change in the intervening timespan, but eventually form tends to mould to function. Or it may happen very quickly; witness the demolition of many 1960s and 1970s housing estates within only a few decades.

It is when part of a city is being re-planned that function becomes a significant driver of form. When a new masterplan is being crafted for an area, then those drivers of human activity come to the fore and have most influence on the outcome. If the city planners, their political masters and the establishment generally believe that it has become necessary to improve the connectivity between city centre and outside world, then new motorways or railways will be driven through the urban landscape, with scant regard to the impact on local communities (Jane Jacob's bug bear). If the city planners and their masters and everyone else deems it essential to build large quantities or new housing as cheaply and quickly as possible, then the likely outcome will be highly repetitive buildings imposed upon the townscape with scant regard to either the context or whether such new housing might be conducive to the creation of new communities rearing children (the legacy of the 1960s). Reflecting back to Chapter 10, each of these scenarios can be construed to be one or other of the identified drivers of urban dynamics being predominant; the values and drivers of different political eras become expressed into and onto the city form accordingly.

The message in this book is that good urban design is fundamentally about balancing those drivers of urban dynamics, no more or less than it is important for a healthy human being to balance his or her day-to-day life. The successful urban designer must be able to transcend scales in time and space, to see how the different influences arising from different types of human activity should be balanced against each other to create an optimum urban form, which generates macro and micro accessibility and enclosure all at once. Focus too much on any one issue, and a solution is doomed to failure. This is no different to the challenges of designing an individual house; if an architect focuses wholly and unerringly on creating the most energy efficient house ever, then he or she is bound to create something that is simply uninhabitable, for example failing to deliver layers of privacy for the inhabitants.

Seeing the evolution of the city over longer timespans enables other perspectives of the relationship between urban form and society to be better appreciated. The twentieth century has witnessed a new human experiment, that of low-density living. It is not the intention of this book to seek to disparage low-density suburbs; there are very clear benefits at a personal level, which arise from access and ownership of a car or cars. The extra private space available in the home and garden, in which to live and bring up a family, the ability to own your own private swimming pool and all manner of other benefits are all fabulous. The mobility to choose where and when to visit what and who is fantastic. But there are inevitable social, as well as environmental, consequences to this form of living.

As take up of the car grew, then there was a very clear iterative process, whereby these benefits accrued to car owners, who could now live beyond the previous limits of the city. Everyone else wanted the same. Young cities in America and Australia and elsewhere rapidly sprawled outwards. At first this was driven by those who wanted to move away from the crowded city centres; but in due course a critical threshold was crossed. As considered in earlier chapters, if you build a city at densities below the 50 dwellings per hectare, then it quickly becomes impossible for the inhabitants to be pedestrian, to walk anywhere other than to the car in their drive. Distances are simply too great to be able to reach anywhere in any reasonable timeframe by foot. Public transport rapidly becomes unviable. This means that the car is no longer a nice to have, but an essential part of everyday life: no car and a person or family becomes completely isolated from the rest of society.

Furthermore, drawing from the thinking earlier in this book, it can be appreciated that, in this low-density urban environment, society becomes an expression of the urban form. The low-density suburb is simply a network of roads. Roads are inherently divisive. The consequence is that there are simply no longer any physical communities. The only communities, which can exist, are a-spatial; they are not geographically grounded. Society inexorably becomes a population of individuals. This is only ameliorated in modern times by an equal growth in communications technology (the telephone, etc.), which allows us to live at low densities without going completely mad, isolated in our own private boxes in their gardens.

In the higher density context, as experienced in times past and which continues to be expressed in both many European cities and those revived city centres in the UK and US, then two very different types of community co-exist within the urban landscape. There are the local spatial communities, which occupy discrete parts of the town (a street or couple of street blocks, a cul-de-sac or the community focussed around a square, or some such) and layered over these locationally fixed communities are all the a-spatial communities of interest – interest groups, hobby groups, sports clubs and so on and so forth. Both types of community have an important part to play in the health of a society.

Where a-spatial communities provide people with mental stimulation and interaction with other like-minded individuals and families, locationally specific communities are important from the perspective of health, well-being and the rearing of children. For the young, such locationally fixed communities provide the social context into which they first experience the city in all its variety beyond containment of the family home. In turn, the community provides the extended eyes and ears for the nervous parents in and around the home. For the elderly, the community provides the eyes and ears to regularly drop by and check that they are okay. For the parent of young children, that local physical community provides the transport to school when a parent or other sibling is ill. The local community is simply the support network for each and every person, the safety net to help out when times are hard (and for each person, family or household, there will always be times when help is needed). The a-spatial community of interest is nice to have; but it doesn't step in and help when the car breaks down …

One of the key things about locational communities is that it is the location, which is frequently the only connector, especially in more established urban neighbourhoods. Neighbours at any urban density will as likely as not be of very different ages and stages of life, with no reason to have any particular commonalities in terms of interests, religion, hobbies, sports, or anything, except location of home. A community of place, therefore, is a very disparate community, including people from all corners of society. Such physical communities consequently instil within individuals in the population a greater sense of community. Where such locational communities disappear, then such sense of diversity of society withers, and with it is lost a sense that society should be supportive of others who may be different to ourselves. Where only communities of interest exist, then individuals in the population become accustomed only ever to dealing with people with commonalities (a common hobby, for example). Society as a consequence is far more splintered into discrete social elements, which have much less if any contact with other communities of interest.

It is well documented that there are significant cultural differences between the US and Europe, in particular with respect to attitudes towards the welfare state. European countries as a whole are far more embracing of the notions of a welfare state, or national health systems and other state sponsored support systems. The US is known to be a society with a far greater individualistic approach. Whilst the American way may have strong historic roots relating to the expansion of society into and across such a vast landscape, it is the twentieth century experiment in low-density living which now enshrines and perpetuates a culture and society which places such priority on individual liberty. Society and urban landscape have become expressions of each other. This is most starkly expressed by viewing the political maps of the American landscape; it is no accident that the Democrats hold sway in the higher density parts of America and the Republicans are more strongly represented in the vast low-density areas. It is simply a direct consequence of the urban worlds in which these populations live.

The Challenge of the Modern Era – How to deal with the Car?

As already recognised several times, the car has been an amazing invention, which has truly liberated individuals and families. In those more developed countries around the world, where mass use of the car has become the norm, human societies have been transformed. Our personal relationship with geographic distance has been equally altered, beyond comprehension. We have, however, perhaps allowed the magic of the car to run away with us, adapting our towns and cities to fit the needs of road users, without much thought for the consequences.

The impact of traffic and heavy road use on the individual (whether as driver, passenger or observer) can be readily appreciated. The car travelling at almost any speed is a dangerous object, able to kill and maim with statistical indiscretion. Much has been done to mitigate these impacts in terms of all aspects of road safety, in car and out. Traffic generates very blatant impacts on individuals in terms of health and nuisance; again much has been done to try to mitigate these affects through improved technology, reducing noise and air pollution. Fuel efficiency is continually improving and alternative technologies, such as electrical vehicles, are waiting their day to come. But the consequence of the car and traffic volumes to society and its communities is something that we have collectively glossed over. The benefit of liberation to the individual is generally seen to far outweigh the disadvantages to the coherence, even existence, of locational communities.

Recognising those noted negativities to the individual associated with the car and traffic, we dream of a world of quiet, clean, ultra-safe road traffic, using yet more technology to generate green electricity in noiseless motors, surrounded by a multitude of electronic sensors to eliminate accidents. What a world that would be! Perhaps it would be desirable to live in the city again, not out in the suburbs. Or alternatively, we could live in the suburbs without conscience, knowing that our chosen lifestyle was without consequence. But, try as we might, no amount of technology can mitigate the social impact of the car.

What, I hear the reader say, about the parallel communications revolution, the phone, the internet and all. Surely they can knit back our communities, allowing communities of place to exist at low-densities, where everyone can live on their own acre of land. If only … From a sociological perspective, a major thing that we are collectively learning about the phone and internet systems is that they accentuate communities of interest. Put a voice of difference into an internet forum and they seemingly attract no end of abuse. The internet does not and, as yet, seemingly cannot create the equivalent of a community of place, one that deals with and embraces difference and diversity and all that goes with that.

In other areas the communications revolution also increases the impacts of the transport revolution. Witness, for example, the damage to our market towns arising from liberal planning policies, which have allowed the creation of out of town shopping areas, leaving once thriving economic hubs with empty hearts. Any chance to resuscitate these towns has been dealt a final blow with the growth of internet shopping. But does it truly matter. We may rue the loss of those dear market towns; do they really have a place in modern society? To answer this question, consider the logical conclusion of the low-density suburb.

In parts of America, the process of social disintegration noted above, where communities of place cease to exist, has become very visibly expressed onto the urban landscape. The logical conclusion of a town, which is made up of roads alone, where streets have ceased to exist, where there are no longer any public spaces in the townscape and where everything has become a private destination, is a town where all residential properties are located within areas which have also become privatised destinations. These are known otherwise as gated communities. Where the gated community differs from normal residential areas is that they effectively demarcate a particular residential area as a membership area. Membership is generally obtained by dint of wealth. But it can be expressed in other ways too, such as foreign enclaves within many middle-eastern or far-eastern cities, or gated communities which only welcome those of a particular religion. These scenarios all point to a dissociation of society into its component parts – into communities of interest rather than communities of locality.

In the market town, which has lost its heart to some out of town food stores, then such dissociation of society is expressed in terms of those who tend to shop at different stores. ASDA customers go only to ASDA and Waitrose clientele visit only Waitrose: never the twain shall meet. All of a sudden a town, which could historically have been considered to have a single community of people, rich and poor, healthy and sick, of all religious creeds, and all age groups all mixing daily and weekly in the market place, is no longer a singular heterogenous community, but instead is two or more divergent homogeneous communities, which rarely meet.

Is this the society we want? The public space at the heart of a town is the mixing ground, where everyone is free to be, to visit, to interact, to socialise and rub shoulders with others, alike or not. The car has been a fabulous, fantastic, liberating invention. But if we give the car priority, when we plan our towns and cities, we will very rapidly find ourselves living in a very different type of society – a society where we only ever interact with people who are the same as ourselves, a society where our only interaction with different people is in the manner of road-rage at the side of a highway. Is this a society we want?

A key challenge for the twenty-first century is to be more rational in our city planning, to appreciate the impact that the automobile can have on our urban landscape and the relationship between built form and social cohesion, and while recognising that private transport is here to stay, making sure that it does not dominate our approach to city planning.

The Next Steps for Urban Design and Urban Planning

The purpose of this book has been to show that it is possible to construct a comprehensive framework for urban design – *a theory of the town*. However, to do so, requires approaching urban design from the perspective of how human beings, individually and collectively, use space, in particular how we cooperate in or compete for the use of space.

The core message from this book is that to understand the structure of the city, then the urban planner must appreciate that form follows function, that the structure of the city is determined by the individual and collective actions and interactions of the inhabitants of the city. By understanding our towns and cities in this way, we can begin to see the similarities between them, what makes a successful urban structure, as opposed to seeing every city as different. As Fernand Braudel appreciated, the city as a concept is ultimately recognisable – *a town is a town wherever it is* (Braudel, 1973) – it is just a matter of asking the right questions.

Appendix A
Urban Vectors for a Cross-road

The core text in Chapter 3 provides a simplified summary of the suggested urban vectors with respect to a road. As discussed in Chapter 3, a true road consists of discrete carriageways, as a minimum one for each direction of travel, to facilitate the flow of traffic. To fully appreciate how the suggested urban vectors would apply to a cross-road the vectors for each carriageway need to be treated separately (Figure A.1). As also mentioned in Chapter 3, this does not apply to pure streets, where there should be no carriageways and where there is no net movement of activity in any particular direction on any part of the highway.

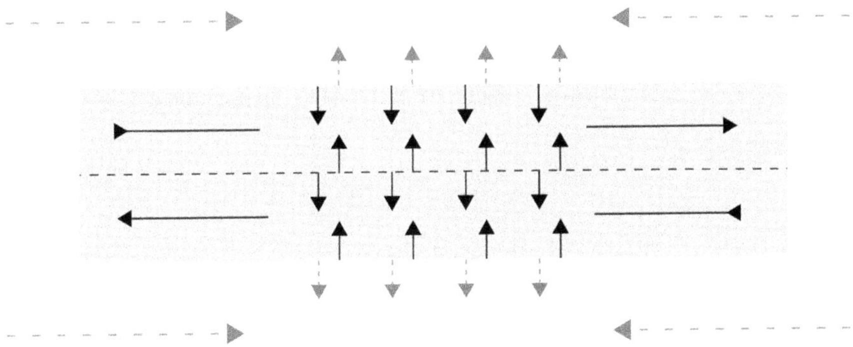

Figure A.1 Full set of vectors for a road

The distinct lanes for a road can be split without affecting the function of the road; and in practice directional lanes often are split apart with a central barrier, sometimes with a very significant gap between the carriageways; or, as per the white lines of British roads, these can be dotted on straight sections and then solid lines on corners to prevent traffic from crossing into oncoming traffic when visibility is reduced. For a proportion of the traffic at a cross-road, the junction does indeed represent a corner.

Where the lanes of a road are separated, then the area between them automatically becomes a *dead zone*, entirely divorced and isolated from the surrounding urban or other landscape (Figure A.2).

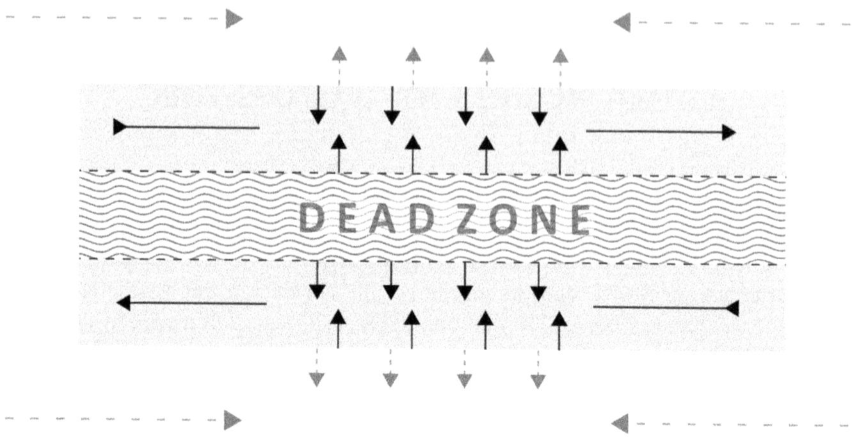

Figure A.2 Urban vectors for a road with separated lanes

In the case of larger cross-roads, especially those where it is desirable to maintain flow of traffic or where there is no clear hierarchy of the roads, then there is a tendency to split the lanes to facilitate entrance to the junction (Figure A.3).

Figure A.3 Temporary separation of lanes to facilitate a junction

And when this is layered onto the other roads forming part of the junction, it becomes apparent why we end up with a roundabout, which is a complete dead zone or *anti-place*, in the centre of the junction (Figure A.4).

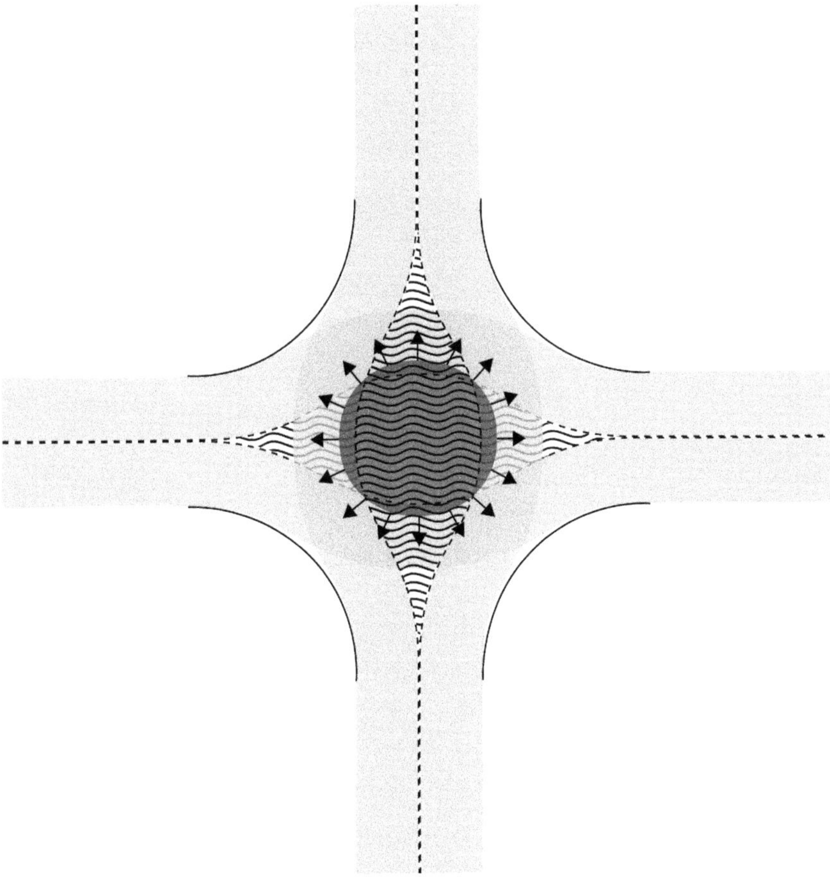

Figure A.4 Roundabout

References

ACPO, 2014. *Secured by Design: New Homes.* [online] Available at: http://www.securedbydesign.com/pdfs/SBDNewHomes2014.pdf [Accessed November 2014].

Alexander, C., 1966. A city is not a tree. *Design.* London: Council of Industrial Design, N° 206.

Appleyard, D., 1969. The environmental quality of city streets: the residents' viewpoint. *Journal of the American Planning Association* 35: 84–101.

Appleyard, D., 1980. Streets Can Kill Cities: Third World Beware!. *Working Paper 336.* Berkeley, CA: Institute of Urban and Regional Development, University of California.

Bryson, B., 2011. *At Home: A Short History of Private Life.* London: Anchor.

Batty, M. and Longley, P., 1994. *Fractal Cities: A Geometry of Form and Function.* London: Elsevier Science.

Burgess, E.W., 1924. The growth of the city: an introduction to a research project. *Publications of the American Sociological Society* 18: 85–97.

Braudel, F., 1973. *Capitalism and Material Life, 1400–1800.* London: HarperCollins.

CABE, 2000. *By Design.* [online] Available at: https://www.gov.uk/government/publications/urban-design-in-the-planning-system-towards-better-practice [Accessed November 2014].

Carmona, M., Heath, T., Oc, T. and Tiesdall, S., 2003. *Public Places: Urban Spaces, The Dimensions of Urban Design.* Architectural Press.

Carmona, M., 2010. *Public Places Urban Spaces: The Dimensions of Urban Design.* London: Taylor & Francis.

Çatalhöyük Research Project, 2013. *Archive Reports.* [online] Available at: http://www.catalhoyuk.com/archive_reports/ [Accessed November 2014].

Coleman, A., 1985. *Utopia on Trial.* London: Shipman.

Daneshpour, A.M., Embi R. and Torabi M., 2012. Privacy in Housing: Effective Variables, 2nd International Conference-Workshop on Sustainable Architecture and Urban Design, Universiti Sains Malaysia, Malaysia, Mar. 2012.

Dead Mall Stories, 2014. [online] Available at: http://www.deadmalls.com/ [Accessed November 2014].

Design for Homes, 2007. *Recommendations for Living at Superdensity.* [online] Available at: http://www.designforhomes.org/wp-content/uploads/2012/03/Superdensity2.pdf [Accessed December 2014].

Dunbar, R., 1992. Neocortex size as a constraint on group size in primates. *Journal of Human Evolution* 20: 469–93.

Dunbar, R., 1993. Coevolution of neocortical size, group size and language in humans. *Behavioral and Brain Sciences* 16 (4): 681–735.
Dunbar, R., 1998. *Grooming, Gossip, and the Evolution of Language*. Harvard University Press.
Energy Savings Trust, 2008. *The Applicability of District Heating for New Dwellings (CE299).* [online] Available at: https://www.b2match.eu/system/low-carbon-heat-showcase-2014/files/District_Heating_for_New_Builds.pdf?1395756062 [Accessed December 2014].
English Partnerships, 2000. *Urban Design Compendium 1*. [online] Available at: http://udc.homesandcommunities.co.uk/sites/default/files/book/udc/community-engagement/1110_udc1_final_artwork_120306-optimized.pdf [Accessed December 2014].
English Partnerships, 2007. *Urban Design Compendium 1*. [online] Available at: http://udc.homesandcommunities.co.uk/sites/default/files/book/udc/community-engagement/1110_udc2_final_artwork_120306-optimized.pdf [Accessed December 2014].
Fat Knowledge: for your fat brain, 2007. *How Much Energy Does an Elevator Use?* [online] Available at: http://fatknowledge.blogspot.co.uk/2007/02/how-much-energy-does-elevator-use.html [Accessed December 2014].
Federal Highway Administration, 2001. *National Household Travel Survey.*
Glaeser, E., 2012. *Triumph of the City*, Penguin.
Greater London Authority, 2010. *Interim London Housing Design Guide.*
Greater London Authority (1), 2011. *Supplementary Planning Guidance for Housing.*
Greater London Authority (2), 2011. *The London Plan.*
Harris, C.D. and Ullman, E.L., 1945. The nature of cities. *Annals of the American Academy of Political and Social Science* 242: 7–17.
Hart, J. and Birkbeck, D., 2013. *New London Vernacular*. [online] Available at: http://www.urbandesignlondon.com/?page_id=7433 [Accessed November 2014].
Hart, J. and Parkhurst, G., 2011. Driven to excess: Impacts of motor vehicles on the quality of life of residents of three streets in Bristol UK. *World Transport Policy & Practice* 17 (2): 12–30.
Homes and Communities Agency, 2010. *Employment Densities Guide, 2nd Edition.*
Hoyt, H., 1939, *The Structure and Growth of Residential Neighborhoods in American Cities*. Washington DC: Federal Housing Administration.
Jacobs, J., 1961. *The Death and Life of Great American Cities*. New York: Random House.
Jacobs, J., 1969. *The Economy of Cities*. New York: Random House.
Kenworthy, J., 2003. Transport Energy Use and Greenhouse Gases in Urban Passenger Transport Systems: A Study of 84 Global Cities. In: *Third Conference of the Regional Government Network for Sustainable Development*. Australia, Fremantle, Notre Dame University, September 17–19, 2003.
Kotkin, J., 2005. *The City: A Global History*, Modern Library Chronicles.

Levitt, D., 2009. *The Housing Design Handbook: A Guide to Good Practice*. London: Routledge.

London Development Agency, 2009. *Working Paper to Support the Outer London Commission: Economic Profile of Key Location in Outer London*. [online] Available at: http://www.london.gov.uk/olc/docs/economic-profile-london.pdf [Accessed November 2014].

Mulholland Research Consultancy, 2003. *Perceptions of Privacy and Density in Housing*. [online] Available at: http://webarchive.nationalarchives.gov.uk/20110118095356/http:/www.cabe.org.uk/files/perceptions-of-privacy-and-density-in-housing.pdf [Accessed November 2014].

Mumford, L., 1961. *The City in History: Its Origins, its Transformations, and its Projects*. Harcourt Inc.

National Statistics (1), 2010. *English Housing Survey of 2010: Homes Report and Annexes*. [online] Available at: https://www.gov.uk/government/statistics/english-housing-survey-2008-to-2009-household-report [Accessed November 2014].

National Statistics (2), 2012. *Commercial and Industrial Floorspace and Rateable Value Statistics (2008)*. [online] Available at: https://www.gov.uk/government/statistical-data-sets/live-tables-on-commercial-and-industrial-floorspace-and-rateable-value-statistics [Accessed November 2014].

National Travel Statistics, 2013. [online] Available at: https://www.gov.uk/government/uploads/system/uploads/attachment_data/file/342160/nts2013–01.pdf [Accessed December 2014].

Newman, O., 1972. *Defensible Space*. New York: Macmillan.

Newman, P. and Kenworthy, J., 1989. *Cities and Automobile Dependence: An International Sourcebook*. Brookfield: Gower.

Newman P. and Kenworthy J., 2006. Urban design to reduce automobile dependence. *Opolis* 2 (1): 35–52.

Noble D. and Smith D., et al., 2012. Feasibility study of geospatial mapping of chronic disease risk to inform public health commissioning, *BMJ Open*, Feb 15, 2 (1): e000711.

ODPM, 2004. *Safer Places – The Planning System and Crime Prevention*. [online] Available at: https://www.gov.uk/government/publications/safer-places-the-planning-system-and-crime-prevention [Accessed November 2014].

Portas, M., 2011. *The Portas Review an Independent Review into the Future of our High Streets*. [online] Available at: https://www.gov.uk/government/uploads/system/attachment_data/file/6292/2081646.pdf [Accessed November 2014].

Pushkarev, B.S. and Zupan J. M., 1997. *Public Transportation and Land Use Policy*. Bloomington: Indiana University Press.

Reader, J., 2005. *Cities*. Vintage Books.

Steemers, K., 2003. Energy and the city: density, buildings and transport. *Energy and Buildings* 35 (1): 3–14.

The Economist, 2014. *Essay – A Planet of Suburbs*. [online] Available at: http://www.economist.com/suburbs [Accessed January 2015].

Thomas, R., (ed.), 2003. *Sustainable Urban Design: An Environmental Approach.* Spon Press.

Triple Pundit: people, planet, profit., 2007. *Ask Pablo: Going Up.* [online] Available at: http://www.triplepundit.com/2007/04/askpablo-going-up/ [Accessed December 2014].

Urban Task Force, 1999. *Towards an Urban Renaissance.* London: Routledge.

van Haaren, R., 2011. *Assessment of Electric Cars' Range Requirements and Usage Patterns based on Driving Behavior Recorded in the National Household Travel Survey of 2009.* [online] Available at: http://www.solarjourneyusa.com/HowFarWeDrive_v1.2.pdf [Accessed December 2014].

Wikipedia, 2014. *Urban Structure.* [online] Available at: http://en.wikipedia.org/wiki/Urban_structure [Accessed December 2014].

Wikipedia, 2014. *Central Place Theory.* [online] Available at: http://en.wikipedia.org/wiki/Central_place_theory [Accessed December 2014].

Wilson, J.Q. and Kelling, G.L., 1982. *Broken Windows: The Police and Neighborhood Safety.* [online] Available at: www.theatlantic.com/politics/crime/windows.htm [Accessed November 2014].

Index

For Product Safety Concerns and Information please contact our EU representative GPSR@taylorandfrancis.com
Taylor & Francis Verlag GmbH, Kaufingerstraße 24, 80331 München, Germany

www.ingramcontent.com/pod-product-compliance
Ingram Content Group UK Ltd.
Pitfield, Milton Keynes, MK11 3LW, UK
UKHW022020180926
475701UK00005B/942
* 9 7 8 0 3 6 7 7 8 7 3 2 5 *